Histamine in Inflammation

Recent Volumes in this Series

Volume 701
OXYGEN TRANSPORT TO TISSUE XXXII
Duane F. Bruley and J.C. LaManna

Volume 702
RNA EXOSOME
Torben Heick Jensen

Volume 703
INFLAMMATION AND RETINAL DISEASE
John D. Lambris and Anthony P. Adamis

Volume 704
TRANSIENT RECEPTOR POTENTIAL CHANNELS
Md. Shahidul Islam

Volume 705
THE MOLECULAR IMMUNOLOGY OF COMPLEX CARBOHYDRATES-3
Albert M. Wu

Volume 706
ADHESION-GPCRs: STRUCTURE TO FUNCTION
Simon Yona and Martin Stacey

Volume 707
HORMONAL AND GENETIC BASIS OF SEXUAL DIFFERENTIATION DISORDERS AND HOT TOPICS IN ENDOCRINOLOGY
Maria I. New and Joe Leigh Simpson

Volume 708
INVERTEBRATE IMMUNITY
Kenneth Söderhäll

Volume 709
HISTAMINE IN INFLAMMATION
Robin L. Thurmond

Histamine in Inflammation

Edited by

Robin L. Thurmond, PhD
Johnson & Johnson Pharmaceutical Research & Development, LLC, San Diego, California, USA

Springer Science+Business Media, LLC

Landes Bioscience

Springer Science+Business Media, LLC
Landes Bioscience

Springer Science+Business Media, LLC, 233 Spring Street, New York, New York 10013, USA
http://www.springer.com

Please address all inquiries to the publishers:
Landes Bioscience, 1806 Rio Grande, Austin, Texas 78701, USA
Phone: 512/ 637 6050; FAX: 512/ 637 6079
http://www.landesbioscience.com

The chapters in this book are available in the Madame Curie Bioscience Database.
http://www.landesbioscience.com/curie

Histamine in Inflammation, edited by Robin L. Thurmond. Landes Bioscience / Springer Science+Business Media, LLC dual imprint / Springer series: Advances in Experimental Medicine and Biology.

ISBN: 978-1-4419-8055-7

While the authors, editors and publisher believe that drug selection and dosage and the specifications and usage of equipment and devices, as set forth in this book, are in accord with current recommendations and practice at the time of publication, they make no warranty, expressed or implied, with respect to material described in this book. In view of the ongoing research, equipment development, changes in governmental regulations and the rapid accumulation of information relating to the biomedical sciences, the reader is urged to carefully review and evaluate the information provided herein.

Library of Congress Cataloging-in-Publication Data

Histamine in inflammation / edited by Robin L. Thurmond.
p. ; cm. -- (Advances in experimental medicine and biology ; v. 709)
Includes bibliographical references and index.
ISBN 978-1-4419-8055-7
1. Histamine--Physiological effect. 2. Histamine--Pathophysiology. 3. Inflammation--Mediators. I. Thurmond, Robin L. (Robin LeRoy), 1965- II. Series: Advances in experimental medicine and biology ; v. 709.
[DNLM: 1. Histamine--physiology. 2. Histamine Antagonists--therapeutic use. 3. Hypersensitivity--physiopathology. 4. Inflammation--physiopathology. 5. Receptors, Histamine--physiology. W1 AD559 v.709 2010 / QU 61]
QP801.H5H575 2010
616'.0473--dc22

2010042469

DEDICATION

This book is dedicated to my wife Tori and my daughters Claire and Grace. It is also dedicated to the memory of two people who were always an inspiration to me, but who unfortunately passed away during the preparation of this work—my grandmother, Georgia Luckett, and one of the pillars of histamine research, Sir James Black.

PREFACE

The year 2010 marks the centennial for the identification of histamine and the first glimpse of its many physiological functions. From these initial findings a rich tapestry of research has uncovered roles for histamine in almost every physiological process with new findings emerging every year. These diverse roles of histamine have made for fertile ground for the discovery of novel therapeutics, and these drugs have been so successful that the term "antihistamine" has entered the common lexicon. This volume is an attempt to give a snapshot in time as to the current understanding of the role of histamine in just one important therapeutic area—inflammation.

The first three chapters provide some background context for the rest of the book starting out with a historical perspective by Figueroa and Shankley. Bongers et al provide an overview of the pharmacology of the four histamine receptors and the chapter by Hiroshi Ohtsu describes how histamine is synthesized as well as the insights derived from mice where this synthesis is disrupted. The next several chapters discuss disease areas where histamine is known to be involved. Chapter 4 by Thomas Taylor-Clark outlines the role of histamine in allergic rhinitis, an area were antihistamines are commonly used. This is also true for ocular allergy as discussed by Ohbayashi et al. Both of these chapters highlight aspects of these conditions that are still not well-controlled and suggest the utility of new antihistamines targeting other histamine receptors. A related conclusion can be seen in the chapter by Dunford and Holgate on the role of histamine in asthma. In this case, however, current antihistamines are not considered to be effective, but the circumstantial evidence for a role of histamine in asthma provides evidence that other histamine receptors may be involved. Antihistamines are also useful for the treatment of some dermatological diseases as discussed by Zuberbier and Maurer. However, the chapter by Buddenkotte et al shows that this cannot be generalized since atopic dermatitis is much like asthma in that histamine has been suspected as being involved, but where current antihistamines are not effective. Traditionally histamine has mainly been associated with allergic reactions, but Schneider et al discuss evidence that histamine may have a broader role in immune function and autoimmune disease. While the majority of other contributions focus on inflammatory conditions, two of the concluding chapters touch on other areas. Nuutinen and Panula discuss the important role of histamine in neurotransmission and for the treatment of various neurological disorders. Falus et al takes the same approach with cell proliferation with an emphasis on malignancy. Finally, the book concludes with the future of antihistamine research and the potential for novel antihistamines targeting newest members of the histamine receptor family—the H_3 and H_4 receptors.

Taken together, I hope that this volume imparts the rich history of histamine research and that it stimulates further interest in uncovering yet to be discovered functions of histamine and the development of new antihistamines for the treatment of human disease.

Robin L. Thurmond, PhD
Johnson & Johnson Pharmaceutical Research & Development, LLC,
San Diego, California, USA

ABOUT THE EDITOR...

ROBIN L. THURMOND, PhD is a Compound Development Team Leader with the Clinical Research group at Johnson & Johnson Pharmaceutical Research & Development in La Jolla, California. Prior to that he was a Research Fellow with the Immunology Drug Discovery group at the same site. He received his BA in Chemistry from the University of Virginia and his PhD in Biochemistry from the University of Arizona. Dr. Thurmond studied membrane biophysics at the University of Arizona and worked on the molecular aspects of rhodopsin function during his postdoctoral training at the Massachusetts Institute of Technology with Dr. Gobind Khorana. He began his career with Johnson & Johnson in 1996 at the RW Johnson Pharmaceutical Research Institute in Raritan, New Jersey, and has been with Johnson & Johnson for over 14 years.

PARTICIPANTS

Pascal Bonaventure
Johnson & Johnson Pharmaceutical Research & Development, LLC
San Diego, California
USA

Gerold Bongers
Leiden/Amsterdam Center for Drug Research (LACDR)
Division of Medicinal Chemistry
Vrije Universiteit Amsterdam
Amsterdam
The Netherlands

Jörg Buddenkotte
Department of Dermatology
Ludwig Boltzmann-Institute for Cell Biology and Immunobiology of the Skin
University Hospital Münster
Münster
Germany

Zsuzsanna Darvas
Department of Genetics, Cell- and Immunobiology
Semmelweis University
Budapest
Hungary

Paul J. Dunford
Department of Immunology
Johnson & Johnson Pharmaceutical Research and Development, LLC
San Diego, California
USA

Michel Dy
Université Paris Descartes
Faculté de Médecine
CNRS UMR8147
Hôpital Necker
Paris
France

Iwan de Esch
Leiden/Amsterdam Center for Drug Research (LACDR)
Division of Medicinal Chemistry
Vrije Universiteit Amsterdam
Amsterdam
The Netherlands

Andras Falus
Department of Genetics, Cell- and Immunobiology
Semmelweis University
Budapest
Hungary
and
Hungarian Academy of Sciences
Research Group for Inflammation Biology and Immunogenomics
Semmelweis University
Budapest
Hungary

Katherine Figueroa
Johnson & Johnson Pharmaceutical Research and Development, LLC
San Diego, California
USA

Ken Fukuda
Department of Ophthalmology
Emory Eye Center
Emory University
Atlanta, Georgia
USA

Stephen T. Holgate
IIR Division
Southampton General Hospital
Southampton
UK

Maria Leite-de-Moraes
Université Paris Descartes
Faculté de Médecine
CNRS UMR8147
Hôpital Necker
Paris
France

Rob Leurs
Leiden/Amsterdam Center for Drug Research (LACDR)
Division of Medicinal Chemistry
Vrije Universiteit Amsterdam
Amsterdam
The Netherlands

Bita Manzouri
Department of Cornea and External Disease
Moorfields Eye Hospital
London, England
UK

Marcus Maurer
Department of Dermatology and Allergy
Allergie-Centrum-Charité
Charité – Universitätsmedizin Berlin
Berlin
Germany

Kei Morohoshi
Department of Ophthalmology
Emory Eye Center
Emory University
Atlanta, Georgia
USA

Saara Nuutinen
Institute of Biomedicine
University of Helsinki
Helsinki
Finland

Masaharu Ohbayashi
Department of Ophthalmology
Emory Eye Center
Emory University
Atlanta, Georgia
USA

Hiroshi Ohtsu
Graduate School of Engineering
Tohoku University
Sendai
Japan

Santa J. Ono
Department of Ophthalmology
Emory Eye Center
Emory University
Atlanta, Georgia
USA

Pertti Panula
Institute of Biomedicine
University of Helsinki
Helsinki
Finland

Zoltán Pós
Department of Genetics, Cell- and Immunobiology
Semmelweis University
Budapest
Hungary
and
Department of Transfusion Medicine
Infectious Disease and Immunogenetics Section
Clinical Center, NIH
Bethesda, Maryland
USA

Elke Schneider
Université Paris Descartes
Faculté de Médecine
CNRS UMR8147
Hôpital Necker
Paris
France

Nigel Shankley
Johnson & Johnson Pharmaceutical Research and Development, LLC
San Diego, California
USA

Martin Steinhoff
Department of Dermatology
Ludwig Boltzmann-Institute for Cell Biology and Immunobiology of the Skin
University Hospital Münster
Münster
Germany
and
Departments of Dermatology and Surgery
University of California San Francisco
San Francisco, California
USA

Thomas Taylor-Clark
Department of Molecular Pharmacology and Physiology
School of Basic Biomedical Sciences
University of South Florida
Tampa, Florida
USA

Robin L. Thurmond
Johnson & Johnson Pharmaceutical Research & Development, LLC
San Diego, California
USA

Fuqu Yu
Johnson & Johnson Pharmaceutical Research & Development, LLC
San Diego, California
USA

Torsten Zuberbier
Department of Dermatology and Allergy
Allergie-Centrum-Charité
Charité – Universitätsmedizin Berlin
Berlin
Germany

CONTENTS

CHAPTER 1

One Hundred Years of Histamine Research

Katherine Figueroa* and Nigel Shankley

Abstract

In this introductory chapter, we revisit some of the landmarks in the history of histamine research. Since histamine was first synthesized (1907) and isolated as a bacterial contaminant of an extract of ergot (1910), the elucidation of its role in health and disease and its molecular mechanism of action have been continuous, reflecting the application of advances in scientific knowledge, technology and therapeutics over the last 100 years.[1] It appears that the research will continue indefinitely as the nature of the problem is inherently fractal. First, there was a single chemical entity, described in terms of state-of-the-art, two-dimensional projections of structures introduced by Fischer in 1891, and an idea that such potent chemicals produced their effects on biological systems as a consequence of an exquisite interaction with a receptive substance, the revolutionary concept of Langley (1905).[2] Today, we recognize four receptor subtypes with multiple activation states and multiple coupling to intracellular effector systems, so that we are no longer able to reliably and in all instances classify compounds interacting with the histamine receptors simply as agonists or antagonists. The complexity is potentially overwhelming, but the promise of value to patients beyond that already provided by the first approved generations of histamine receptor blockers is a compelling driver.

Introduction

In the Middle Ages, bewitchment would sometimes take the form of St Anthony's fire, a syndrome characterized by convulsions, diarrhea, gangrene and hallucinations. Rather than a questionable 'gift' of the local witch, in the 1850s this syndrome was associated with the ingestion of rye and other cereals infected with a fungi, Claviceps purperea. It was determined that the over-wintering body of the fungi, termed ergot, with the appearance of their host grain, was responsible for these physiological and mental symptoms allowing the syndrome to be renamed as the ergot poisoning still observed today. The therapeutic value of preparations of ergot was recognized centuries earlier by midwives seeking to control post-partum hemorrhage on account of the uterine contractions they induced. The elucidation of the diverse pharmacologically active ingredients of ergot began with the development of extraction and isolation chemistry techniques in the 19th Century. For example, in 1875, Tanret isolated the crystalline alkaloid ergotinine, a powerful and fast-acting uterine contractile agent, allowing for safe administration with measured dosing.[3]

In 1904, Henry Dale aged 29, following his education at Cambridge University under the mentorship of John Langley among other leading physiologists, was appointed as a researcher at the Wellcome Physiological Research Laboratories in South London. According to Dale's memoirs[1]

*Corresponding Author: Katherine Figueroa—Johnson & Johnson Pharmaceutical Research and Development L.L.C., Merryfield Row, San Diego, California, USA.
Email: kfiguero@its.jnj.com

Histamine in Inflammation, edited by Robin L. Thurmond.

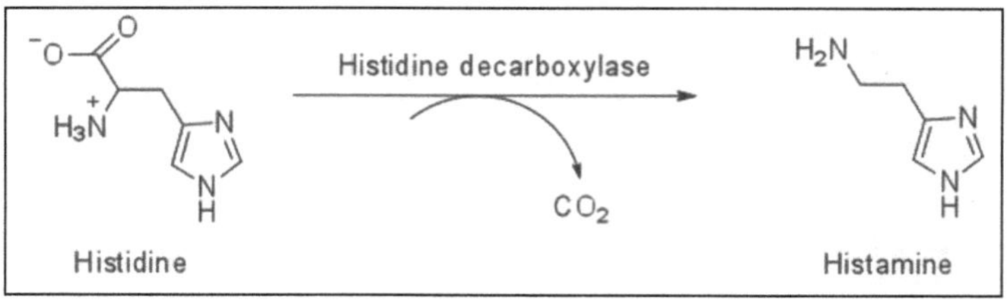

Figure 1. Synthesis and distribution of histamine. Histamine can be released from a variety of cells including neurons, enterochromaffin-like cells and mast cells. Histamine is generated by decarboxylation of histidine by the enzyme histidine decarboxylase.

(1953), Henry Wellcome requested that he "make an attempt to clear up the problem of ergot; the pharmacy, pharmacology and therapeutics of that drug being then in a state of obvious confusion". Together with George Barger, a chemist and former Cambridge associate, who had already prepared a number of compounds from ergot, Dale began a systematic pharmacological evaluation of each compound. Working with a standardized, although decaying and contaminated, "Liquid Extract of Ergot" Barger & Dale (1910)[4] reported the isolation and identification of β-aminoethylimidazole, the product of the decarboxylation of the amino-acid histidine (Fig. 1). According to Dale, the logical short name, histamine, was withheld for a while due to trademark infringement disputes. The chemical synthesis of histamine had first been described three years earlier by Windhaus and Vogt (1907),[5] providing a method for Dale and colleagues to obtain a supply of pure material for their subsequent physiological studies.

Exploring the Physiological Effects of Histamine

The early exploration of the biological effects of histamine was predominantly performed by observation and measurement in mammalian experimental systems. Histamine was described as having powerful contractile effects on a variety of smooth muscles including the uterus, airways and certain blood vessels but was also found to be a secretogogue stimulating gastric acid secretion. In addition Dale and Laidlaw (1910)[6] made the link between the phenomenon of anaphylaxis and histamine when they recognized the similarity of the response, and subsequent death of sensitized guinea-pigs, to a second challenge with an antigen to that produced by histamine. In 1927, Dale's laboratory, in collaboration with Best, went on to demonstrate that histamine was an endogenous ligand by showing that extracts obtained from liver and lung produced similar responses to chemically synthesized histamine in isolated uteri of guinea-pigs and anesthetized cats.[7]

The First Antihistamines

In 1936, at the Pasteur Institute, Bovet and Staub used compounds synthesized by Fourneau to show that certain stimulatory actions of histamine could be prevented.[8] The first reported 'antihistamine' was 933F (piperoxan), which in the isolated guinea-pig ileum could block the effect of histamine. Another compound 929F could prevent death due to histamine injection in guinea-pigs but had too many side effects to be considered for clinical use. Members of a later series, which included phenbenzamine (RP 2339, Antergan), were able to prevent anaphylaxis-induced bronchospasm in guinea pig as well as many actions of histamine in various other species. These compounds possessed a lower level of side effects enabling mepyramine (RP2786, pyrilamine) to be marketed as Neoantergan in 1944 for the treatment of allergies.[8,9] Although the early pharmacological analysis of antihistamines was largely qualitative, it soon became apparent that not all of the physiological effects of histamine were blocked by this new class of agents.

Pharmacological Definition of Histamine H_1 and H_2 Receptors

In the 1940s, evidence was presented first by Wells et al (1945) and subsequently by Folkow et al (1948), suggesting that the antihistamines prevented the actions of histamine as a consequence of competitive inhibition at a cell surface receptor.[10,11] The fact that specific substances, like benadryl, only partially reversed histamine-induced vasodilation in the cat led to the conclusion that histamine-induced vasodilation was mediated by more than one type of receptor. Around this time, Schild, following on from the pioneering work of Clark in 1927 and Gaddum in 1937, was refining analytical methods for the use of competitive receptor antagonists as quantitative tools for receptor classification.[12]

In 1960, Trendelenburg[13] used these methods to show the pA_2 values obtained for pyrilamine, when blocking the actions of histamine to increase rate in the isolated mouse right atria and smooth muscle contractions in the isolated guinea-pig ileum, were significantly different and so concluded that the histamine receptors in the two tissues were pharmacologically distinct.[13] This initial observation was followed by the first classification of histamine receptors by Ash and Schild.[14] It was determined that low concentrations of specific histamine antagonists inhibited receptors in the guinea-pig ileum and bronchi but did not produce the same response in rat uterine contraction assays. The receptor expressed in the guinea-pig ileum and bronchi was classified as H_1. It was noted that classification of a second histamine receptor would require selective antagonists targeted towards to those expressed in the uterus, right atrium and stomach. This was solved by studies commenced at Smith, Kline and French in 1964.

James Black, following his successful discovery of β-blockers working with medicinal chemist John Stephenson within the ICI Pharmaceutical Division, relocated to Smith, Kline and French laboratories and established a new partnership with Robin Ganellin.[15] They took the same approach followed for the discovery of the first β-blockers (pronethalol and propranolol) namely, by making analogues of the parent hormone (adrenaline) and β-selective agonists (isoprenaline) and partial agonists (dichloroisoprenaline). 4-methylhistamine was found to express selective agonist activity for what would later be termed the H_2 receptor. Subsequently, *N*-(α)-guanylhistamine, was identified as a selective partial agonist and further modification led to the prototype, selective, competitive antagonists, burimamide and metiamide.[9,16] Burimamide not only inhibited gastric acid secretion in the dog but also was shown to inhibit histamine pressor responses in the cat addressing the complex cardiovascular pharmacology of histamine first described by Dale 60 years previously. The clinical development of metiamide was terminated due to toxicity in preclinical testing but the replacement compound, cimetidine (Tagamet®), was subsequently marketed for the treatment of heart burn and peptic ulcers.

Pharmacological Definition of the Histamine H_3 Receptor

The presence of a third histamine receptor was first clearly described by Arrang et al, in 1983.[17] Just as the existence of the H_2 receptor was first postulated by observation that not all actions of histamine were blocked by the H_1-receptor antagonists, it was the unexpected pharmacological behavior of the compounds previously classified as selective H_2-receptor agonists and antagonists that characterized the H_3-receptor. Their paper described the inhibition of $[^3H]$-histamine release from depolarized slices of rat cerebral cortex through a histamine-stimulated presynaptic receptor. Burimamide, the prototype low potency H_2-receptor antagonist ($pK_B \sim 5$) and impromidine, previously classified as a potent, selective H_2-receptor partial agonist, both behaved as potent competitive antagonists with burimamide expressing ~300-fold higher affinity for the new autoreceptor. It was later shown that the expression of the third histamine receptor was not confined to the central nervous system. In 1987, Trzeciakowski described presynaptic inhibition of myenteric nerve stimulated contractions of the guinea-pig ileum by a class of receptors that pharmacologically resembled the H_3-receptor described earlier by Arrang et al.[18] In the same year, the first potent and selective H_3-receptor compounds were described, *R*-(α)-methylhistamine and thioperamide, an agonist and antagonist, respectively, which remain as standard reference compounds.[17] Although no histamine H_3-receptor compounds have been approved as drugs, there are several centrally-acting antagonists currently in clinical trials for the treatment of disorders such as Alzheimer's disease, Attention-Deficit-Hyperactivity-Disorder (ADHD) and narcolepsy.

Pharmacological Definition of the Histamine H_4 Receptor

History repeated itself in the early 1990s as the fourth histamine was also characterized initially using quantitative pharmacological methods and existing compounds classified in terms of their potency and affinity for agonist and antagonist activity, respectively, at the H_1-, H_2- and H_3-receptors. Raible et al (1996) first discovered that histamine-stimulated increases in intracellular calcium in eosinophils were blocked by thioperamide, but not by pyrilamine or cimetidine, suggesting that the effects of histamine may be H_3-receptor mediated.[19] However, they found that the H_3-receptor selective agonist, *R*-(α)-methylhistamine, was over an order of magnitude less potent than histamine, which was not consistent with the characterization of H_3-receptors in other tissues and prompted them to perform a more detailed pharmacological analysis using a broader range of compounds. They concluded that the eosinophil histamine receptor was novel. The two H_3-receptor antagonists, thioperamide and impromidine expressed affinity estimates similar to their H_3-receptor values. However, histamine was more potent than, not only *R*-(α)-methylhistamine, but also *N*-(α)-methylhistamine (previously characterized as an agonist at H_1-, H_2- and H_3-receptors with highest potency at the H_3-receptor) and dimaprit which behaved as a low potency partial agonist (previously characterized as an agonist at H_2-receptors and antagonist at H_3-receptors). With the subsequent cloning and expression of the histamine H_4-receptor (see below), highly selective H_4–receptor antagonists have been described over the last few years. Given the expression of the receptor on key immune cells these ligands have been advocated as having potential therapeutic value in auto-immune and allergic disease although to date no clinical trial data have been reported.[20]

The Molecular Biological Characterization of Histamine Receptors

By 1990, although there was a significant amount of literature pertaining to the function of the three known histamine receptors, there was little information about the molecular structure of these receptors. The first histamine receptor to be cloned was the canine H_2-receptor in 1991 by Gantz et al at the University of Michigan, Ann Arbor.[21] Their work came after other G protein-linked receptors had been cloned with the amino-acid sequence suggesting a rhodopsin-like seven transmembrane structure. In order to clone the canine H_2-receptor Gantz and colleagues employed the method developed by Libert et al.[22] The process involved the use of degenerate primers and polymerase chain reactions (PCR) to generate partial cDNA sequences from gastric parietal cell mRNA; the PCR derived clones were then used to probe canine genomic libraries. Colo-320 DM cells transfected with the vector containing the canine H_2-receptor clone exhibited an increase in intracellular cAMP when challenged with histamine and the concentration-response curve to histamine could be shifted to the right by prior incubation with cimetidine. The selective H_2-receptor antagonist also inhibited [methyl-^{3}H]tiotidine binding in the H_2-receptor expressing cells. Later that same year functional human and rat H_2-receptors were cloned and expressed using similar techniques.[23,24]

The bovine H_1-receptor was the first one of the subtype to be cloned.[25] In an alternative method to genomic library screening, the method employed by Yamashita and collegues involved adrenal medullar RNA fractions injected into Xenopus oocytes for assessment of Ca^{2+}-dependent Cl^- currents, the presence of which and its susceptibility to mepyramine, indicated that the RNA fraction injected could transcribe a functional H_1-receptor. Northern blot analysis showed that the cloned H_1-receptor had highest expression in the lung and small intestine. Functional human and rat H_1-receptors were cloned and expressed two years later.[26,27]

Within a decade of cloning the first histamine receptor, cloning of the human H_3-receptor was reported in the literature.[28,29] Lovenberg et al described the screening of a human thalamus library using a GPCR fragment from an orphan library database, GPR97. A full-length clone encoding a putative GPCR was isolated and, although it showed high homology to the acetycholine muscarinic M_2-receptor, presented pharmacology that was indistinguishable from the histamine H_3-receptor. Expression analysis revealed high expression in many brain areas, confirming histamine's role as a neurotransmitter modulator as first revealed 16 years earlier by Arrang et al in 1983.[17] When the receptor was cloned from multiple species it became evident that there are significant differences in the H_3-receptor pharmacological profile between species.[30,31]

Soon after the confirmation of the H_3-receptor sequence, the histamine H_4-receptor was cloned through the use of genomic database homology screening, which had also been the means for successful determination of the H_3-receptor sequence.[32-35] A partial clone, GPR 105, had significant homology to the newly cloned H_3-receptor. When GPR105 was transfected into SK-N-MC cells it expressed high specific binding for [^{3}H]-histamine but not for H_1- or H_2-receptor specific radiolabels. Moreover, binding to GPR105 was inhibited by a panel of H_3-receptor antagonists with an unusual rank order of affinity, once again indicating the existence of a novel histamine receptor. In contrast to the H_3-receptor, which is predominantly expressed in the central nervous system, the H_4-receptor was found to be primarily expressed in bone marrow and eosinophils.

Alternative Intracellular Signaling of Histamine Receptors

Following the pioneering work of Earl Sutherland and colleagues in 1959 to reveal cyclic adenine monophosphate (cAMP) as an intracellular messenger, Karppanen and Westermann (1973) showed that histamine dose-dependently stimulated the production of cyclic AMP in the gastric mucosa of guinea pigs.[36,37] This effect could be selectively inhibited by H_2- but not H_1-receptor antagonists, suggesting that the H_2-receptor was positively coupled to adenylate cyclase. In the same year, Lichtenstein and Gillespie also showed positive modulation by histamine of intracellular cAMP levels in human leukocytes.[38] Through H_2-receptor selective inhibition they identified a negative feedback mechanism whereby histamine inhibits its own release from human leukocytes via the H_2-receptor. For the H_1-receptor, it was later shown that histamine provokes turnover of inositol phospholipids in guinea-pig and human airway epithelial cells via a G protein-dependent mechanism.[39]

The H_1 through H_4 histamine receptors have now been shown to couple to $G_{q/11}$, G_s and $G_{i/o}$ G-proteins, respectively.[24,28-30,40] In addition to increases in intracellular inositol phosphates upon activation of the H_1-receptor it was frequently noted that significant increases in cAMP were observed upon receptor activation. In 2005, Maruko et al described that upon activation of the H_1-receptor the $\beta\gamma$-dimer released after separation from the $G\alpha_{q/11}$ protein, stimulated the increases in cAMP, which could be inhibited by co-transfection of $G\alpha_s$-protein C-terminal peptides.[41] This result suggests the histamine family of receptors should be added to the growing number of GPCRs that may have differential effects on cellular function and organ physiology through possible 'ligand directed signaling'.

A summary of cloned histamine receptors, their signaling pathways, distribution and selective ligands is presented in Table 1.

Generations of Histamine Targeted Genetically Modified Mice

Identification of the genes encoding each of the four histamine receptor subtypes enabled the generation of genetically modified mice lacking each of the receptors H_1 through H_4. The first histamine receptor 'knock-out' mouse was generated targeting the H_1-receptor.[42] Both physiological and neurological testing showed the lack of functional H_1-receptors produced similar results to the administration of selective H_1-receptor antagonists. An increase in brain serotonin levels was noted as a compensatory response in the H_1-receptor knock-out mice that could account for the behavioral changes observed. The generation of an H_2-receptor knock-out mouse by Kobayashi et al in 2000 confirmed previous evidence that the H_2-receptor is absolutely required for normal function of parietal cell acid secretion and cellular homeostasis of the gastric mucosa.[43] H_2-receptor knock-out mice present normal gastric basal pH but display gastric hypertrophy and increased circulating gastrin levels. Neither histamine nor gastrin stimulate gastric acid secretion in the H_2-receptor knock-out mice. Compensatory mechanisms for this change in physiology were shown to include an increased cholinergic stimulus to the parietal cells. Prior to the generation of H_3-receptor knock-out mice, the function of this predominantly presynaptic autoreceptor was evaluated using the selective antagonist

Table 1. Histamine receptor subtype summary

	H_1	H_2	H_3	H_4
Amino acid sequence	NP_000852 HRH_1: 487 amino acids	NP_071640 HRH_2: 359 amino acids	NP_009163 HRH_3: 455 amino acids	NP_067637 HRH_4: 390 amino acids
Coupling	G_q/G_{11} family to phospholipase C stimulation	G_s family to adenylate cyclase stimulation	$G_{i/o}$ family to adenylate cyclase inhibition	$G_{i/o}$ family to intracellular calcium increases
Distribution	CNS, airway smooth muscle, gastrointestinal tract, cardiovascular system, lymphocytes, endothelial cells, genitourinary system, adrenal medulla	stomach, vascular smooth muscle, CNS, cardiovascular system, neutrophils, uterus	CNS, cardiovascular system, lungs, endothelial cells, peripheral nerves	Eosinophils, bone marrow and leukocytes
Selective agonists	HTMT 2-pyridylethylamine	amthamine, dimaprit	imetit methyl-histamine	Clobenpropit VUF 8430
Selective antagonists	mepyramine, fexofenadine, diphenhydramine	cimetidine, ranitidine	clobenpropit, ROS 234	JNJ7777120 JNJ10191584 thioperamide
Radiolabel	[^{3}H]-pyrilamine	[^{125}I]-aminopotentidine	[^{3}I]-iodoproxyfan	[^{3}H]-JNJ7777120

thioperamide.[44] The generation of the H_3-receptor knock-out mice in 2002 by Toyota et al[43,45] provided an alternative method to evaluate the receptors role in behavior modulation.[45] The H_3-receptor knock-out mice showed normal circadian rhythmicity but did not respond to the wake-inducing properties of thioperamide. The H_3-receptor deficient mice also highlighted the role the H_3-receptor plays in the modulation of other neurotransmitter systems in the brain such as the noradrenergic and cholinergic control of attention and memory functions, respectively. The link between H_3-receptor modulation and dopaminergic control has yet to be fully characterized. The last histamine receptor knock-out mouse to be generated and evaluated was of the H_4-receptor subtype.[46] Initial studies using the H_4-receptor deficient mice showed the receptor was important for histamine mediated mast-cell chemotaxis. Overall, the use of histamine receptor-deficient mice, in combination with specific receptor antagonists, continues to allow clarification of histamine's role in various pathologies and guide potential therapeutic value of selective antagonists.

Conclusion

As will be detailed in the following chapters of this book, investigation into the modulation of histamine signaling continues and is still a significant area of pharmacological research and drug discovery and development. From the use of H_1-receptor antagonists for the treatment of urticaria

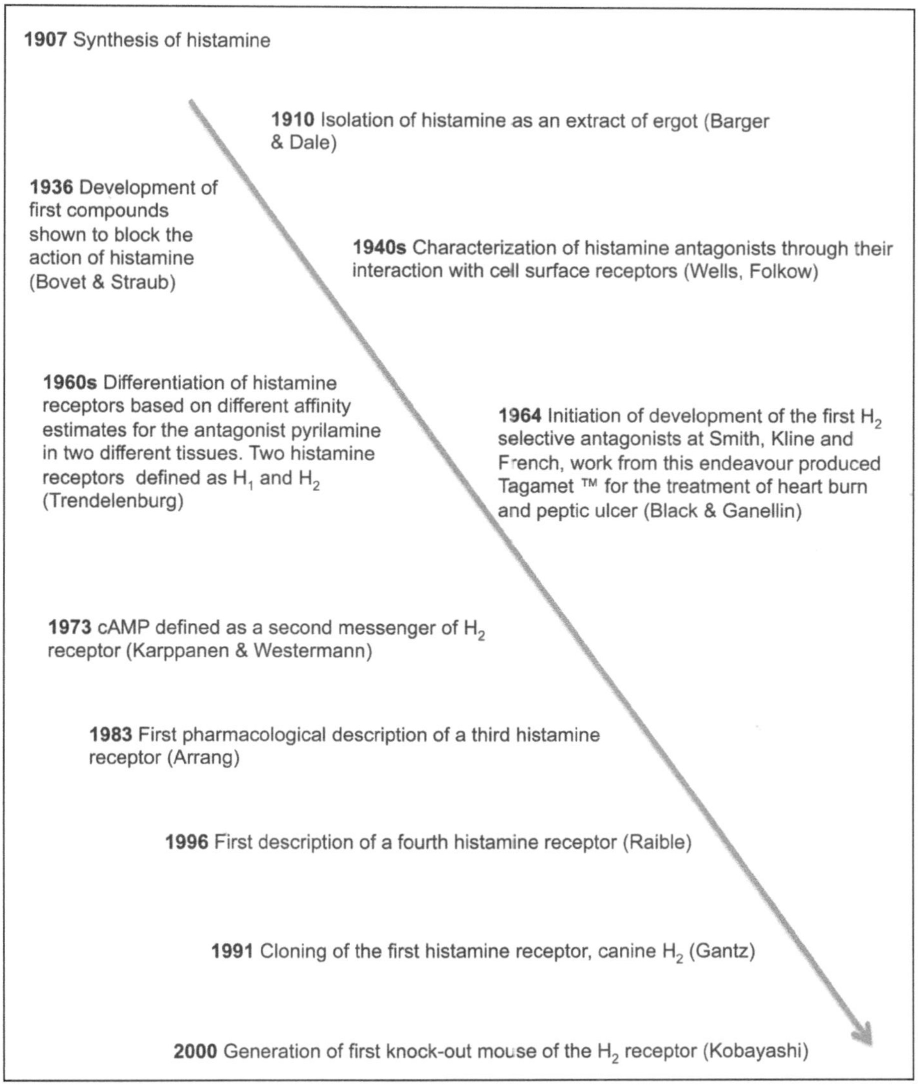

Figure 2. Historic milestones during one hundred years of histamine research.

since the 1950s and to the discovery of the H_2 antagonists for gastric acid hypersecretion disorders, the area of histamine pharmacology now moves on to both neurological and immunological disorders (see Fig. 2 for a diagrammatic synopsis of the one hundred years of histamine research). Expression of the H_3-receptor in the brain led to research into the control of body weight and appetite, narcolepsy and Alzheimer's. The identification of H_4 receptors has seen a number of new classes of selective antagonist which may be of value in diseases such as allergic rhinitis and atopic dermatitis. One thing seems certain; Dale's contaminant of ergot will continue to occupy researchers during the next century.

Acknowledgement

We would like to thank Drs. Timothy Lovenberg and Nicholas Carruthers for their helpful review and comments.

References

1. Dale HH. Adventures in Physiology. Pergamon Press, Ltd., 1953.
2. Langley JN. On the reaction of cells and of nerve endings to certain poisons, chiefly as regards the reaction of striated muscle to nicotine and to curate. J Physiol 1905; 33:374-413.
3. Delvincourt R. Vie et oeuvre de Charles Tanret. Thesis d'Universite Paris 1980; 5:548.
4. Barger G, Dale HH. 4-b-Aminoethylglyoxaline (b-Aminazolylethylamine) and the other active principles of ergot. J Chem Soc Trans 1910; 97:2592-5.
5. Windhaus A, Vogt W. Synthese des imidazolyl-athylamins. Ber Dtsch Chem Ges 1907; 40:3691-3695
6. Dale HH, Laidlaw PP. The physiological action of b-iminazolyethylamine. J Physiol 1910; 16(5):318-344.
7. Best CH, Dale HH, Dudley HW et al. The nature of the vaso-dilator constituents of certain tissue extracts. J Physiol 1927; 62(4):397-417.
8. Bovet D. Introduction to antihistamine agents and antergan derivative. Ann NY Acad Sci 1950; 50(9):1089-126.
9. Parsons E, Ganellin CG. Histamine and its receptors. Br J Pharmacol 2006; 147:S127-S135.
10. Wells J, Morris H, HB. B et al. Observations on the nature of antagonism of histamine by β–dimethylaminoethyl bezhydryl ether. J Pharmacol Exp Ther 1945; 85:122-128.
11. Folkow B, Heger K, Kahlson G. Observations on reactive hyperaemia as related to histamine on drugs antagonising vasodilation induced by histamine and on vasodilator properties of adenosine triphosphate. Acta Physiol Scan 1948; 15:264-278.
12. Aranlakshan O, Schild H. Some quantitative uses of drug antagonists. Br J Pharmacol Chemother 1959; 14:48-58.
13. Trendelenburg U. The action of histamine and 5-hydroxytryptamine on isolated mammalian atria. J Pharmacol Exp Ther 1960; 130(4):450-460.
14. Ash A, Schild H. Receptors mediating some actions of histamine. Br J Pharmacol Chemother 1966; 27(2):427-439.
15. Black J. A personal view of pharmacology. Annu Rev Pharmacol Toxicol 1996; 36:1-33.
16. Black J, Duncan W, Durant C et al. Definition and antagonism of histamine H_2-receptors. Nature 1972; 236:385-390.
17. Arrang J, Garbarg M, Schwartz J. Auto-inhibition of brain histamine release mediated by a novel class (H_3) of histamine receptor. Nature 1983; 302(5911):832-837.
18. Trzeciakowski JP. Inhibition of guinea pig ileum contractions mediated by a class of histamine receptor resembling the H_3 subtype. J Pharmacol Exp Ther 1987; 243(3):874-880.
19. Raible D, Lenahan T, Fayvilevich Y et al. Pharmacologic characterization of a novel histamine receptor on human eosinophils. Am J Respir Crit Care Med 1996; 149(6):1506-1511.
20. Thurmond RL, Gelfand EW, Dunford PJ. The role of histamine H_1 and H_4 receptors in allergic inflammation: the search for new antihistamines. Nat Rev Drug Discov 2008; 7(1):41-53.
21. Gantz I, Schaffer M, DelValle J et al. Molecular cloning of a gene encoding the histamine H2 receptor. Proc Natl Acad Sci USA 1991; 88(2):429-433.
22. Libert F, Parmentier M, Lefort A et al. Selective amplification and cloning of four new members of the G protein-coupled receptor family. Science 1989; 244(4904):569-572.
23. Ganz PR et al. Molecular cloning of the human histamine H_2 receptor. Biochem Biophys Res Comm 1991; 178:1386
24. Ruat M, Traiffort E, Arrang J-M et al. Cloning and tissue expression of a rat histamine H_2-receptor gene. Biochem Biophys Res Commun 1991; 179(3):1470-1478.
25. Yamashita M, Fukui H, Sugama K et al. Expression cloning of a cDNA encoding the bovine histamine H_1 receptor. Proc Natl Acad Sci USA 1991; 88(24):11515-11519.
26. Debacker MD, Gommeren W, Moereels H et al. Genomic cloning, heterologous expression and pharmacological characterization of a human histamine H_1 receptor. Biochem Biophys Res Commun 1993; 197(3):1601-1608.
27. Fujimoto K, Horio Y, Sugama K et al. Genomic cloning of the rat histamine H_1 receptor. Biochem Biophys Res Commun 1993; 190(1):294-301.
28. Lovenberg TW, Roland BL, Wilson SJ et al. Accelerated communication: cloning and functional expression of the human histamine H_3 receptor. Mol Pharmacol 1999; 55(6):1101-1107.
29. Morse KL, Behan J, Laz TM et al. Cloning and characterization of a novel human histamine receptor. J Pharmacol Exp Ther 2001; 296(3):1058-1066.
30. Lovenberg TW, Pyati J, Chang H et al. Cloning of rat histamine H_3 receptor reveals distinct species pharmacological profiles. J Pharmacol Exp Ther 2000; 293(3):771-778.
31. Chen J, Liu C, Lovenberg TW. Molecular and pharmacological characterization of the mouse histamine H_3 receptor. Eur J Pharmacol 2003; 467(1-3):57-65.
32. Nakamura T, Itadani H, Hidaka Y et al. Molecular cloning and characterization of a new human histamine receptor, HH4R. Biochem Biophys Res Commun 2000; 279(2):615-620.

33. Nguyen T, Shapiro DA, George SR et al. Discovery of a novel member of the histamine receptor family. Mol Pharmacol 2001; 59(3):427-433.
34. Oda T, Morikawa N, Saito Y et al. Molecular cloning and characterization of a novel type of histamine receptor preferentially expressed in leukocytes. J Biol Chem 2000; 275(47):36781-36786.
35. Liu C, Ma X-J, Jiang X et al. Cloning and pharmacological characterization of a fourth histamine receptor (H_4) expressed in bone marrow. Mol Pharmacol 2001; 59(3):420-426.
36. Rall TW, Sutherland EW. Formation of a cyclic adenine ribonucleotide by tissue particles. J Biol Chem 1958; 232(2):1065-1076.
37. Karppanen H, Westermann E. Increased production of cyclic AMP in gastric tissue by stimulation of histamine2 (H_2)-receptors. Naunyn Schmiedebergs Arch Pharmacol 1973; 279(1):83-87.
38. Gillespie E, Lichtenstein L. Pharmacologic control of IgE-mediated histamine release from human leukocytes. Int Arch Allergy Appl Immunol 1973; 45(1):95-97.
39. Li H, Choe N, Wright D et al. Histamine provokes turnover of inositol phospholipids in guinea pig and human airway epithelial cells via an H_1-receptor/G protein-dependent mechanism. Am J Respir Cell Mol Biol 1995; 12(4):416-424.
40. Gantz I, Munzert G, Tashiro T et al. Molecular cloning of the human histamine H_2 receptor. Biochem Biophys Res Commun 1991; 178(3):1386-1392.
41. Maruko T, Nakahara T, Sakamoto K et al. Involvement of the $\beta\gamma$ subunits of G proteins in the cAMP response induced by stimulation of the histamine H_1 receptor. Naunyn Schmiedebergs Arch Pharmacol 2005; 372(2):153-159.
42. Yanai K, Son LZ, Endou M et al. Behavioural characterization and amounts of brain monoamines and their metabolites in mice lacking histamine H_1 receptors. Neuroscience 1998; 87(2):479-487.
43. Kobayashi T, Tonai S, Ishihara Y et al. Abnormal functional and morphological regulation of the gastric mucosa in histamine H_2 receptor-deficient mice. J Clin Invest 2000; 105(12):1741-1749.
44. Clark EA, Hill SJ. Sensitivity of histamine H_3 receptor agonist-stimulated [35S]GTP[gamma][S]binding to pertussis toxin. Eur J Pharmacol 1996; 296(2):223-225.
45. Toyota H, Dugovic C, Koehl M et al. Behavioral characterization of mice lacking histamine H_3 receptors. Mol Pharmacol 2002; 62(2):389-397.
46. Hofstra CL, Desai PJ, Thurmond RL et al. Histamine H_4 receptor mediates chemotaxis and calcium mobilization of mast cells. J Pharmacol Exp Ther 2003; 305(3):1212-1221.

Chapter 2

Molecular Pharmacology of the Four Histamine Receptors

Gerold Bongers, Iwan de Esch and Rob Leurs*

Abstract

Histamine and its receptors have been (and are still today) very fruitful topics for pharmacological and medicinal chemistry studies. In this chapter we review the various selective ligands that are available for the four different histamine receptors and we describe the main molecular pharmacological aspects of each of the receptor subtypes.

The Discovery of the Four Histamine Receptors: An Historical Overview

The biological effects of histamine were observed early by Dale and Laidlaw (1910). Injection of the biogenic amine produced similar effects as in many allergic reactions.[1] As early as 1937, the first evidence for a histamine receptor was provided by Bovet and Staub, who discovered the first antihistamine thymoxidiethylamine, that was capable of preventing anaphylactic shock in animals.[2] The discovery by Ash and Schild in 1966 that antihistamines, like mepyramine, could block certain pharmacological actions of histamine on symptoms of allergic reactions, but not the effects on the gastric acid secretion led the hypothesis that there were at least two subtypes of histamine receptors.[3] This was further corroborated by the finding that burimamide selectively antagonized the histamine mediated effects on the gastric acid secretion.[4] The histamine mediated auto-inhibition of brain histamine release was shown to by mediated by a third class of histamine receptors that could be pharmacologically differentiated from the heretofore known histamine H_1 receptor (H_1R) and histamine H_2 receptor (H_2R).[5] The histamine H_3 receptor (H_3R) was definitely confirmed by the first selective and potent H_3R antagonists thioperamide.[6] The last member of the histamine receptor family was originally cloned as an orphan receptor, but based on its high sequence homology to the H_3R was found to respond to histamine and confirmed to be a fourth histamine receptor, the histamine H_4 receptor (H_4R).[7-11]

The four histamine receptors are all membrane bound proteins that belong to the superfamily of the G-protein coupled receptors (GPCRs) and more precisely to the biogenic amine receptors in the rhodopsin-family. GPCRs convert diverse stimuli like odors, photons, neurotransmitters (including biogenic amines), hormones, peptides and proteases, via guanine nucleotide-binding proteins (G-proteins) into intracellular responses. GPCRs are characterized by seven alpha helical transmembrane (TM) domains and are found in eukaryotes, including yeast, plants, choanoflagellates and animals. They are involved in numerous physiological processes like smell, taste, vision, behavior and mood, regulation of the immune system and autonomic nervous system transmission. GPCRs are considered attractive drug targets by the pharmaceutical industry, because they are involved in the regulation of almost every major mammalian physiological process and are

*Corresponding Author: Rob Leurs—Leiden/Amsterdam Center for Drug Research (LACDR), Division of Medicinal Chemistry, Faculty of Sciences, VU University, Amsterdam, De Boelelaan 1083, 1081 HV, Amsterdam, The Netherlands. Email: leurs@few.vu.nl

Histamine in Inflammation, edited by Robin L. Thurmond.

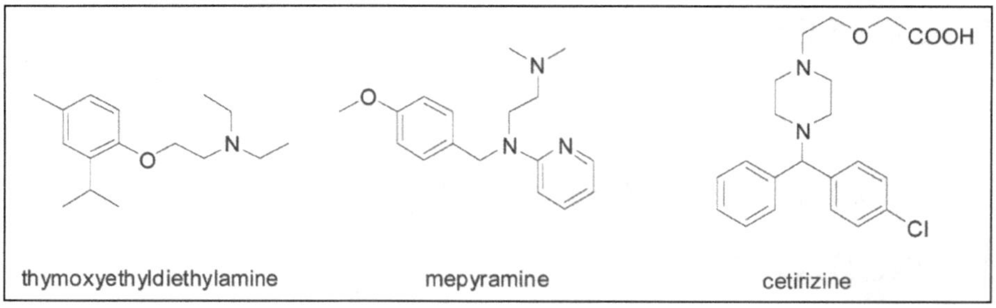

Figure 1. Chemical structures of selected H_1R antagonists.

readily accessible to drugs due to their localization on the cell surface. In fact, 30% of all drugs on the market are targeting GPCRs, among these are several block-buster drugs like clopidogrel (Plavix®), cimetidine (Tagamet®), Fexofenadine hydrochloride (Allegra®), quetiapine (Seroquel®) and metoprolol (Lopressor® or Seloken®).[12,13] Recently, considerable progress has been made in the purification and crystallization of several members of the rhodopsin class of GPCRs,[14-16] likely further enhancing the success of GPCR based drug discovery.

Histamine Receptors, Signal Transduction and Their Ligands

In the next paragraphs we will discuss various molecular pharmacological aspects of the four different receptor subtypes, including the availability of selective subtype selective agonists and antagonists.

The Histamine H_1 Receptor and Its Ligands

The histamine H_1 receptor (H_1R) is found mainly on smooth muscle cells, endothelium and in the CNS. Its physiological role includes e.g., vasodilatation, bronchoconstriction, modulation of endothelial barrier function (responsible for hives), pain and itching due to insect stings. The antagonists for the H_1R, commonly known as antihistamines, are successfully used for the treatment of allergic rhinitis and skin irritations.[17] Following the first antihistamine, thymoxyethyldiethylamine (Fig. 1), the related ethylenediamines (e.g., mepyramine) were the first clinically used H_1R antagonists. Like the other first generation H_1R antagonists, the use of mepyramine however suffers from sedation as a side effect. Actually, these compounds are now used in many sleeping-aid preparations. Second generation antagonists for the H_1R, e.g., the piperazine cetirizine, have a reduced occurrence of adverse drug reactions due to a decreased brain penetration and increased H_1R selectivity.[18]

Some selective H_1R agonists have recently been developed as well, but are not used therapeutically. Modification of the imidazole moiety of histamine has been the most successful approach for obtaining selective H_1 agonists (Fig. 2). The presence of the tautomeric N^{π}-N^{τ} system of the imidazole ring is

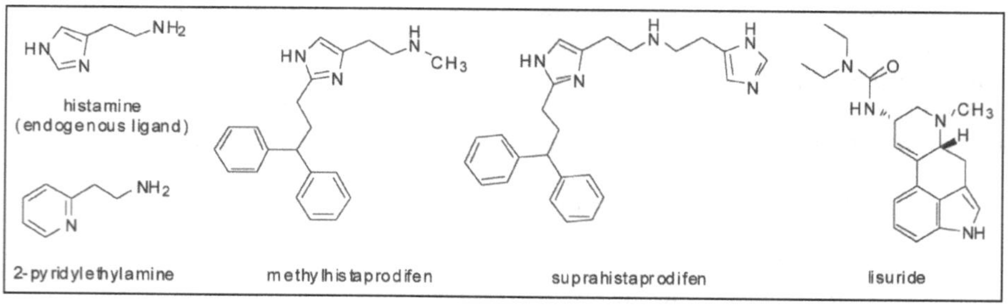

Figure 2. Chemical structures of selected H_1R agonists.

not obligatory, as reflected by the selective, but weak H_1 agonists 2-pyridylethylamine. Substitution of the imidazole ring at the 2-position leads to relatively selective H_1 agonists. Schunack and colleagues developed a series of H_1R selective histaprodifens.[19-21] A further increase in H_1R agonist potency was obtained by a bivalent ligand approach. Suprahistaprodifen, a dimer of histaprodifen and histamine is currently one of the most potent H_1R agonists available.[22,23] Surprisingly, recent high throughput screening of CNS-active drugs at the H_1R has identified the non-imidazole ergot derivative lisuride as another high affinity H_1R agonist.[24]

The bovine H_1R cDNA was cloned from a cDNA library of bovine adrenal medulla and was the first H_1R gene to be cloned,[25] soon to be followed by other species, including the human H_1R.[26-28] The human H_1R gene is an intron-less gene that is located on chromosome 3p25 and encodes for a 487 amino acid GPCR with a long third intracellular loop (IL3).[29]

The H_1R predominantly couples to $G\alpha_{q/11}$-proteins[30] leading to the activation of PLC and subsequent release of the second messengers IP_3 and DAG followed by the activation of PKC and the release of $[Ca^{2+}]_i$. Additionally, the H_1R has been shown to constitutively increase IP_3 levels[31] and to activate the nuclear factor κB (NF-κB),[32] a transcription factor involved in inflammation and cancer. Remarkably, the H_1R-mediated constitutive activation of NF-κB is primarily mediated through G-protein βγ-subunits, whereas both $G\alpha_{q/11}$-proteins and βγ-subunits are required for the H_1R agonists mediated NF-κB activation.[32] All the clinically used H_1R antagonists, in fact act as inverse agonists inhibiting the constitutive activation of the H_1R.

The Histamine H_2 Receptor and Its Ligands

The histamine H_2 receptor (H_2R) is located in a variety of tissues including brain, gastric cells and cardiac tissue.[17] H_2Rs are involved in the gastric acid secretion and therefore antagonists of the H_2R are used in the treatment of peptic ulcers. The first H_2R antagonist, burimamide (Fig. 3),[4] was not very potent and actually was not very specific for the H_2R either. With the discovery of the H_3R and H_4R we now know that burimamide has a higher affinity for the H_3R and H_4R.[5,33] Further development within the class of H_2R antagonists led to the discovery of cimetidine (Tagamet®) by Smith, Kline and French and ranitidine (Zantac®) by GlaxoSmithKline. These H_2R antagonists have been widely used in the clinical treatment of peptic ulcers and have become major blockbusters. Nowadays it has become apparent that gastric ulcers can effectively be cured by a proton-pump inhibitor in combination with antibiotics when an infection with *H. pylori* is found.[34-36]

A first step towards a selective H_2R agonist was made with the discovery of dimaprit (Fig. 4), which was found in a quest for isothiourea-based H_2R antagonists. Dimaprit is an H_2R agonist that is almost as active as histamine at the H_2R, but hardly displays any H_1R agonism. Later it was found that dimaprit is also a moderate H_3R antagonist and a moderate H_4R agonist. Using dimaprit as a template, amthamine (2-amino-5-(2-aminoethyl)-4-methylthiazole) was designed as a rigid dimaprit analogue (Fig. 4).[37] Amthamine combines a high H_2R selectivity with a potency, which is slightly higher compared to histamine, both in vitro and in vivo.

The H_2R gene was the first gene of the histamine receptor family to be cloned. By using degenerate oligonucleotide primers based on the known homology between GPCRs and subsequent polymerase chain reaction (PCR) on canine gastric parietal cell cDNA Gantz and coworkers cloned the canine H_2R.[38] High homology of the various H_2R facilitated cloning of the H_2R in other species, including the human H_2R gene.[39] The human H_2R gene is an intron-less gene

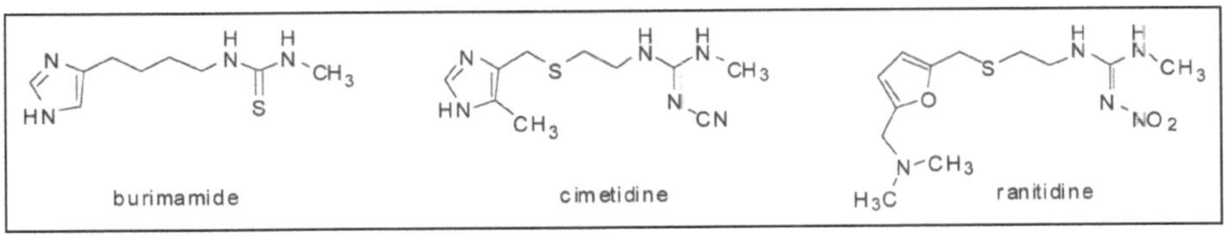

Figure 3. Chemical structures of H_2R antagonists.

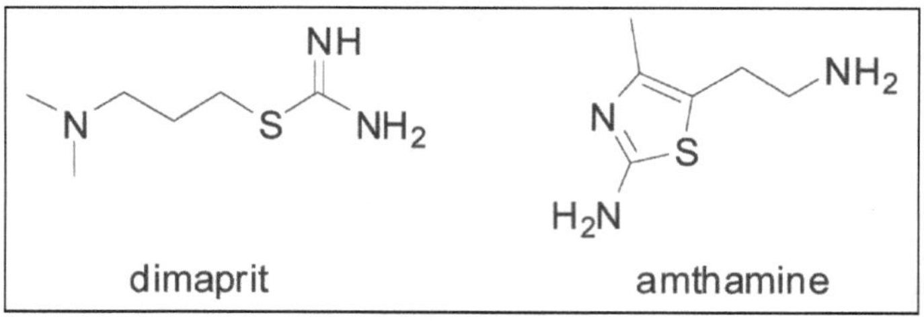

Figure 4. Chemical structures of H_2R agonists.

located on chromosome 5q35 encoding for a protein of 358 amino acids. Compared to the other histamine receptors the H_2R has a short IL3 and a longer C-terminal tail. The H_2R predominantly couples to $G\alpha_s$-proteins and subsequently leads an increase in intracellular cAMP and the activation PKA. Selective immunoprecipitation of activated G proteins labeled with [α-^{32}P]GTP azidoanilide revealed, in addition to $G\alpha_s$-protein coupling, specific coupling to $G\alpha_q$-proteins as well. Stimulation of recombinantly expressed H_2R in COS-7 cells indeed resulted in an increase in intracellular inositol 3-phosphate (IP_3) and as well as an increase in cAMP.[40] Similar to the H_1R, the H_2R was found to display constitutive activity as well,[41] which led to the reclassification of heretofore known (and clinically important) antagonists (like cimetidine and ranitidine) as inverse agonists. Burimamide was found to be neutral antagonist for the rat H_2R,[41] but acts as a weak partial agonist on the human H_2R.[42]

The Histamine H_3 Receptor and Its Ligands

The histamine H_3 receptor (H_3R) is predominantly expressed in the CNS and to a lesser extent in the peripheral nervous system.[17] On histaminergic neurons in the CNS the H_3R acts as an presynaptic autoreceptor inhibiting the release and synthesis of histamine.[5] On nonhistaminergic neurons in mammalian brain, the H_3R functions as a heteroreceptor inhibiting the release of various important neurotransmitters like serotonin, noradrenalin, acetylcholine and dopamine.[17] Besides neuronal expression, peripheral inhibitory effects of H_3R activation on neurotransmission have been shown to occur in the cardiovascular system, gastrointestinal tract and the airways.[43-46]

The H_3R has been an attractive drug target for both academia and the pharmaceutical industry.[47-51] The H_3R is expressed in brain regions that are critical for cognition (cortex and hippocampus), sleep and homeostatic regulation (hypothalamus).[52] Moreover, the H_3R acts as a heteroreceptor modulating the release of several important neurotransmitters that are involved in processes like cognition, mood and sensory gating.[53-55] In addition, the H_3R acts as an autoreceptor regulating the release and synthesis of histamine, a neurotransmitter that plays a role in vigilance, attention, impulsivity and feeding/weight regulation.[17,56] Therefore, antagonists for the H_3R are currently under investigation in several therapeutic areas including sleep disorders, energy homeostasis and cognitive disorders.[57,58]

The first potent H_3R ligands, e.g., thioperamide and clobenpropit (Fig. 5)[6,59] were based on the structure of histamine and therefore imidazole-based. However, development of H_3R specific antagonists by pharmaceutical companies like GlaxoSmithKline (e.g., GSK-189254), Abbott (e.g., A-423579), Johnson and Johnson, Schering-Plough, Pfizer, UCB Pharma, Merck, Banyu, Eli Lilly, Sanofi-Synthelabo and Roche focused on non-imidazole compounds,[57,60] in order to limit potential drug-drug interactions via the interaction with the cytochrome P450 isoenzymes.[61,62]

At the H_3 receptor, histamine itself is a highly active agonist. Methylation of the α-carbon atom of histamine's ethylamine sidechain leads to R-α-methylhistamine, with a highly reduced activity at both the H_1R—and H_2R and potent agonist activity at the H_3R.[63] For potent H_3 agonism, the amine function of histamine can be incorporated in ring structures. For example, immepip (Fig. 6) is a potent H_3 agonist that is effective in vitro and in vivo. Although

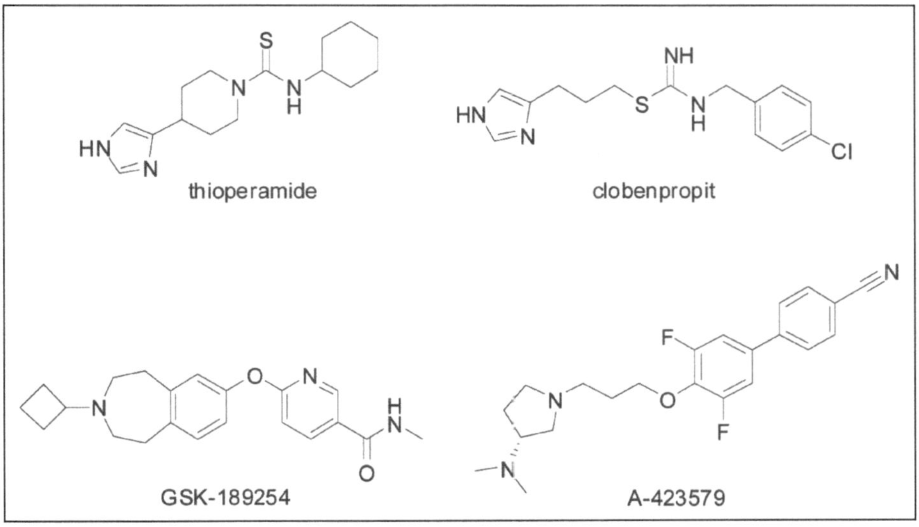

Figure 5. Chemical structures of H_3R antagonists.

immepip and R-α-methylhistamine have previously been used as reference ligands to study the H_3R, both of them have considerable activity for the recently discovered H_4R. Therefore, new potent and selective H_3R agonists have been developed, most notably immethridine ($pEC_{50} = 9.8$; 300 fold selectivity over the H_4R) and methimepip ($pEC_{50} = 9.5$; >10000 fold selectivity over the H_4R).[64]

It was not until the turn of last century before the human H_3R cDNA was identified by Lovenberg and his coworkers at Johnson and Johnson in 1999.[65] Earlier efforts to clone the H_3R gene by homology screening on the basis of the earlier elucidated H_1R and H_2R genes all failed. In search for novel GPCRs in commercial genome databases, an orphan GPCR with homology to the M_2 muscarinic acetylcholine receptor was identified. Full pharmacological characterization of this new aminergic GPCR identified this protein as the histamine H_3R. Cloning of the H_3R genes of other species, including rat, guinea pig and mouse, soon followed and important H_3 receptor species differences have been identified.[66] The H_3R mRNA undergoes extensive alternative splicing, resulting in many H_3 receptor isoforms that have different signaling properties and expression profiles.[50,67] Moreover, the H_3R displays particularly high constitutive activity, which can also be observed in vivo, leading to a reclassification of existing ligands into agonists, neutral antagonists and inverse agonists.

The H_3R signals via $G\alpha_{i/o}$ proteins as shown by the pertussis toxin sensitive stimulation of [^{35}S]-GTPγS binding in rat cortical membranes.[68] The inhibition of adenylyl cyclase after stimulation of the H_3R results in lowering of cellular cAMP levels and modulation of CREB (cAMP responsive element-binding protein) dependent gene transcription.[69-75] Moreover, the H_3R effectively couples to the stimulation of MAPK [67] and the Akt-GSK3-β axis.[76]

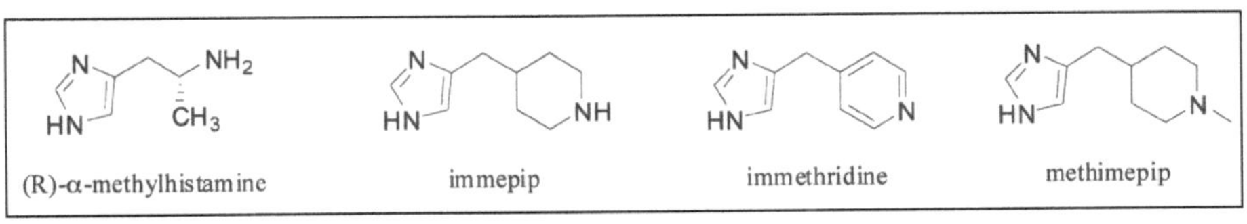

Figure 6. Chemical structures of H_3R agonists.

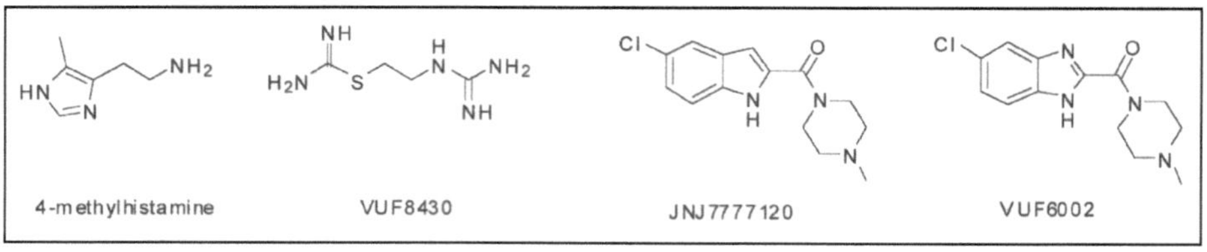

Figure 7. Chemical structures of H_4R ligands.

The Histamine H_4 Receptor and Its Ligands

The histamine H_4 receptor (H_4R) has a relatively low CNS expression and is highly expressed in peripheral blood leukocytes and mast cells, suggesting a role for the H_4R in inflammatory and immune responses.[77] At this moment lot of interest is focused on the potential of the H_4R as drug target in inflammatory conditions (e.g., allergic asthma) and itch. With the discovery of the H_4R and its initial pharmacological characterization,[7-11] it became immediately clear that many imidazole-containing H_3R ligands show a high affinity for the H_4R as well.[78] This is probably due to the high homology (68%) in the transmembrane regions of the H_3R and H_4R.[8,9,33,79] Classical H_3R ligands like the H_3R agonists immepip and imetit and the H_3R inverse agonist clobenpropit were shown to be potent high affinity agonists on the H_4R,[78] whereas thioperamide turns out to be a high affinity inverse agonist for the H_4R.[33] The first potent and H_4R selective agonists, 4-methylhistamine[80] and VUF8430[78] and inverse agonists, JNJ 7777120[81] and its benzimidazole derivative VUF600[2,82,83] have now been developed (Fig. 7). These specific H_4R ligands, together with the availability of H_3R specific ligands,[84] will help to delineate the roles of the H_4R in vivo.

The gene that encodes for the human H_4R is located on chromosome 18q11.2 and contains three exons encoding for a 390 amino acid protein that has a 31% homology to the human H_3R.[85] Similarity in gene organization between the H_3R and the H_4R might indicate the possibility of H_4R isoforms, however so far no 7-TM H_4R isoforms have been published.[7,9,11,86] Like the H_3R, the H_4R couples to $G\alpha_{i/o}$-proteins, subsequently leading to an inhibition of cAMP accumulation and the subsequent PKA dependent inhibition of the cAMP responsive element-binding protein (CREB).[10] Furthermore, activation H_4R has been shown to lead to a $G\alpha_{i/o}$-protein dependent phosphorylation of MAPK in HEK293 cells[87,88] and mobilization of $[Ca^{2+}]_i$ in mast cells endogenously expressing the H_4R and in L1.2 cells that recombinantly express the H_4R.[87] The H_4R mediated mobilization of $[Ca^{2+}]_i$ in mast cells is both $G\alpha_{i/o}$-protein and PLC dependent as shown by the use of PTX and the phospholipase C inhibitor U73122.[65]

Conclusion

After the first successful period of histamine receptor pharmacology and the blockbuster success of the histamine H_1R and H_2R antagonists, the histamine research area is now having a firm revival. With the cloning of the genes of the H_3R and H_4R worldwide significant attention is paid to the potential therapeutic use of ligands acting at these two "new" family members. It is anticipated that in the coming years the first clinical results with recently developed H_3R and H_4R antagonists will be made public.

References

1. Dale HH, Laidlaw PP. J Physiol 1910; 41:318-344.
2. Bovet D, Staub AM. Comp rend Soc Biol 1937; (124):547.
3. Ash AS, Schild HO. Receptors mediating some actions of histamine. Br J Pharmacol Chemother 1966; 27(2):427-439.
4. Black JW, Duncan WA, Durant CJ et al. Definition and antagonism of histamine H_2-receptors. Nature 1972; 236(5347):385-390.
5. Arrang JM, Garbarg M, Schwartz JC. Auto-inhibition of brain histamine release mediated by a novel class (H_3) of histamine receptor. Nature 1983; 302(5911):832-837.

6. Arrang JM, Garbarg M, Lancelot JC et al. Highly potent and selective ligands for histamine H_3-receptors. Nature 1987; 327(6118):117-123.
7. Zhu Y, Michalovich D, Wu H et al. Cloning, expression and pharmacological characterization of a novel human histamine receptor. Mol Pharmacol 2001; 59(3):434-441.
8. Nguyen T, Shapiro DA, George SR et al. Discovery of a novel member of the histamine receptor family. Mol Pharmacol 2001; 59(3):427-433.
9. Liu C, Ma X, Jiang X et al. Cloning and pharmacological characterization of a fourth histamine receptor (H(4)) expressed in bone marrow. Mol Pharmacol 2001; 59(3):420-426.
10. Morse KL, Behan J, Laz TM et al. Cloning and characterization of a novel human histamine receptor. J Pharmacol Exp Ther 2001; 296(3):1058-1066.
11. Oda T, Morikawa N, Saito Y et al. Molecular cloning and characterization of a novel type of histamine receptor preferentially expressed in leukocytes. J Biol Chem 2000; 275(47):36781-36786.
12. Wise A, Gearing K, Rees S. Target validation of G-protein coupled receptors. Drug Discov Today 2002; 7(4):235-246.
13. Service RF. Surviving the blockbuster syndrome. Science 2004; 303(5665):1796-1799.
14. Cherezov V, Rosenbaum D, Hanson MA et al. High-resolution crystal structure of an engineered human beta2-adrenergic G protein-coupled receptor. Science 2007; 318(5854):1258-1265.
15. Rosenbaum D, Cherezov V, Hanson MA et al. GPCR engineering yields high-resolution structural insights into beta2-adrenergic receptor function. Science 2007; 318(5854):1266-1273.
16. Rasmussen S, Choi H, Rosenbaum D et al. Crystal structure of the human beta2 adrenergic G-protein-coupled receptor. Nature 2007; 450(7168):383-387.
17. Hill SJ, Ganellin CR, Timmerman H et al. International Union of Pharmacology. XIII. Classification of histamine receptors. Pharmacol Rev 1997; 49(3):253-278.
18. Kay GG. The effects of antihistamines on cognition and performance. J Allergy Clin Immunol 2000; 105(6 Pt 2):S622-S627.
19. Elz S, Kramer K, Leschke C et al. Ring-substituted histaprodifen analogues as partial agonists for histamine $H_{(1)}$ receptors: synthesis and structure-activity relationships. Eur J Med Chem 2000; 35(1):41-52.
20. Elz S, Kramer K, Pertz HH et al. Histaprodifens: synthesis, pharmacological in vitro evaluation and molecular modeling of a new class of highly active and selective histamine $H_{(1)}$-receptor agonists. J Med Chem 2000; 43(6):1071-1084.
21. Malinowska B, Piszcz J, Schlicker E et al. Histaprodifen, methylhistaprodifen and dimethylhistaprodifen are potent H_1-receptor agonists in the pithed and in the anaesthetized rat. Naunyn Schmiedebergs Arch Pharmacol 1999; 359(1):11-16.
22. Menghin S, Pertz HH, Kramer K et al. N(alpha)-imidazolylalkyl and pyridylalkyl derivatives of histaprodifen: synthesis and in vitro evaluation of highly potent histamine $H_{(1)}$-receptor agonists. J Med Chem 2003; 46(25):5458-5470.
23. Čarman-Kržan M, Bavec A, Zorko M et al. Molecular characterization of specific H_1-receptor agonists histaprodifen and its Nα-substituted analogues on bovine aortic H_1-receptors. Naunyn Schmiedebergs Arch Pharmacol 2003; 367(5):538-546.
24. Bakker RA, Weiner DM, ter Laak T et al. 8R-lisuride is a potent stereospecific histamine H_1-receptor partial agonist. Mol Pharmacol 2004; 65(3):538-549.
25. Yamashita M, Fukui H, Sugama K et al. Expression cloning of a cDNA encoding the bovine histamine H_1 receptor. Proc Natl Acad Sci U S A 1991; 88(24):11515-11519.
26. De Backer MD, Gommeren W, Moereels H et al. Genomic cloning, heterologous expression and pharmacological characterization of a human histamine H_1 receptor. Biochem Biophys Res Commun 1993; 197(3):1601-1608.
27. Fukui H, Fujimoto K, Mizuguchi H et al. Molecular cloning of the human histamine H_1 receptor gene. Biochem Biophys Res Commun 1994; 201(2):894-901.
28. Moguilevsky N, Varsalona F, Noyer M et al. Stable expression of human H_1-histamine-receptor cDNA in Chinese hamster ovary cells. Pharmacological characterisation of the protein, tissue distribution of messenger RNA and chromosomal localisation of the gene. Eur J Biochem/FEBS 1994; 224(2):489-495.
29. De Backer MD, Loonen I, Verhasselt P et al. Structure of the human histamine H_1 receptor gene. Biochem J 1998; 335(Pt 3):663-670.
30. Leopoldt D, Harteneck C, Nurnberg B. G proteins endogenously expressed in Sf 9 cells: interactions with mammalian histamine receptors. Naunyn Schmiedebergs Arch Pharmacol 1997; 356(2):216-224.
31. Bakker RA, Wieland K, Timmerman H et al. Constitutive activity of the histamine $H_{(1)}$ receptor reveals inverse agonism of histamine $H_{(1)}$ receptor antagonists. Eur J Pharmacol 2000; 387(1):R5-R7.
32. Bakker RA, Schoonus SB, Smit MJ et al. Histamine $H_{(1)}$-receptor activation of nuclear factor-kappa B: roles for G beta gamma- and G alpha(q/11)-subunits in constitutive and agonist-mediated signaling. Mol Pharmacol 2001; 60(5):1133-1142.

33. Lim HD, van Rijn RM, Ling P et al. Evaluation of histamine H_1-, H_2- and H_3-receptor ligands at the human histamine H_4 receptor: identification of 4-methylhistamine as the first potent and selective H_4 receptor agonist. J Pharmacol Exp Ther 2005; 314(3):1310-1321.
34. Deakin M, Williams JG. Histamine H_2-receptor antagonists in peptic ulcer disease. Efficacy in healing peptic ulcers. Drugs 1992; 44(5):709-719.
35. Freston JW. Management of peptic ulcers: emerging issues. World J Surg 2000; 24(3):250-255.
36. Penston JG. Review article: clinical aspects of Helicobacter pylori eradication therapy in peptic ulcer disease. Aliment Pharmacol Ther 1996; 10(4):469-486.
37. Eriks JC, van der Goot H, Sterk GJ et al. Histamine H_2-receptor agonists. Synthesis, in vitro pharmacology and qualitative structure-activity relationships of substituted 4- and 5-(2-aminoethyl)thiazoles. J Med Chem 1992; 35(17):3239-3246.
38. Gantz I, Schaffer M, DelValle J et al. Molecular cloning of a gene encoding the histamine H_2 receptor. Proc Natl Acad Sci U S A 1991; 88(2):429-433.
39. Gantz I, Munzert G, Tashiro T et al. Molecular cloning of the human histamine H_2 receptor. Biochem Biophys Res Commun 1991; 178(3):1386-1392.
40. Kuhn B, Schmid A, Harteneck C et al. G proteins of the Gq family couple the H_2 histamine receptor to phospholipase C. Mol Endocrinol 1996; 10(12):1697-1707.
41. Smit MJ, Leurs R, Alewijnse AE et al. Inverse agonism of histamine H_2 antagonist accounts for upregulation of spontaneously active histamine H_2 receptors. Proc Natl Acad Sci U S A 1996; 93(13):6802-6807.
42. Alewijnse AE, Smit MJ, Hoffmann M et al. Constitutive activity and structural instability of the wild-type human H_2 receptor. J Neurochem 1998; 71(2):799-807.
43. Schwartz JC, Arrang JM, Garbarg M et al. Plenary lecture. A third histamine receptor subtype: characterisation, localisation and functions of the H_3-receptor. Agents Actions 1990; 30(1-2):13-23.
44. Bertaccini G, Coruzzi G. An update on histamine H_3 receptors and gastrointestinal functions. Dig Dis Sci 1995; 40(9):2052-2063.
45. Malinowska B, Godlewski G, Schlicker E. Histamine H_3 receptors—general characterization and their function in the cardiovascular system. J Physiol Pharmacol 1998; 49(2):191-211.
46. Delaunois A, Gustin P, Garbarg M et al. Modulation of acetylcholine, capsaicin and substance P effects by histamine H_3 receptors in isolated perfused rabbit lungs. Eur J Pharmacol 1995; 277(2-3):243-250.
47. Alguacil LF, Perez-Garcia C. Histamine H_3 receptor: a potential drug target for the treatment of central nervous system disorders. Curr Drug Targets 2003; 2(5):303-313.
48. Bonaventure P, Letavic M, Dugovic C et al. Histamine H_3 receptor antagonists: from target identification to drug leads. Biochem Pharmacol 2007; 73(8):1084-1096.
49. Hancock AA. The challenge of drug discovery of a GPCR target: analysis of preclinical pharmacology of histamine H_3 antagonists/inverse agonists. Biochem Pharmacol 2006; 71(8):1103-1113.
50. Leurs R, Bakker RA, Timmerman H et al. The histamine H_3 receptor: from gene cloning to H_3 receptor drugs. Nat Rev Drug Discov 2005; 4(2):107-120.
51. Leurs R, Timmerman H. The histamine H_3-receptor: a target for developing new drugs. Progress in drug research. Fortschr Arzneimittelforsch 1992; 39:127-165.
52. Hancock AA, Fox GB. Perspectives on cognitive domains, H_3 receptor ligands and neurological disease. Expert Opin Investig Drugs 2004; 13(10):1237-1248.
53. Caulfield MP, Birdsall NJ. International Union of Pharmacology. XVII. Classification of muscarinic acetylcholine receptors. Pharmacol Rev 1998; 50(2):279-290.
54. Missale C, Nash SR, Robinson SW et al. Dopamine receptors: from structure to function. Physiol Rev 1998; 78(1):189-225.
55. Schlicker E, Malinowska B, Kathmann M et al. Modulation of neurotransmitter release via histamine H_3 heteroreceptors. Fundam Clin Pharmacol 1994; 8(2):128-137.
56. Schwartz JC, Arrang JM, Garbarg M et al. Histaminergic transmission in the mammalian brain. Physiol Rev 1991; 71(1):1-51.
57. Wijtmans M, Leurs R, de Esch I. Histamine H_3 receptor ligands break ground in a remarkable plethora of therapeutic areas. Expert Opin Investig Drugs 2007; 16(7):967-985.
58. Passani MB, Lin JS, Hancock A et al. The histamine H_3 receptor as a novel therapeutic target for cognitive and sleep disorders. Trends Pharmacol Sci 2004; 25(12):618-625.
59. van der Goot H, Schepers MJP, Sterk GJ et al. Isothiourea analogues of histamine as potent agonists or antagonists of the histamine H_3-receptor. Eur J Med Chem 1992; 27(5):511-517.
60. Cowart M, Altenbach R, Black L et al. Medicinal chemistry and biological properties of non-imidazole histamine H_3 antagonists. Mini Rev Med Chem 2004; 4(9):979-992.
61. LaBella FS, Queen G, Glavin G et al. H_3 receptor antagonist, thioperamide, inhibits adrenal steroidogenesis and histamine binding to adrenocortical microsomes and binds to cytochrome P450. Br J Pharmacol 1992; 107(1):161-164.

62. Yang R, Hey JA, Aslanian R et al. Coordination of histamine H_3 receptor antagonists with human adrenal cytochrome P450 enzymes. Pharmacology 2002; 66(3):128-135.
63. Arrang JM, Garbarg M, Lancelot JC et al. Highly potent and selective ligands for a new class H_3 of histamine receptor. Invest Radiol 1988; 23 Suppl 1:S130-S132.
64. Kitbunnadaj R, Zuiderveld OP, Christophe B et al. Identification of 4-(1H-imidazol-4(5)-ylmethyl) pyridine (immethridine) as a novel, potent and highly selective histamine $H_{(3)}$ receptor agonist. J Med Chem 2004; 47(10):2414-2417.
65. Lovenberg TW, Roland BL, Wilson SJ et al. Cloning and functional expression of the human histamine H_3 receptor. Mol Pharmacol 1999; 55(6):1101-1107.
66. Hancock AA, Esbenshade TA, Krueger KM et al. Genetic and pharmacological aspects of histamine H_3 receptor heterogeneity. Life Sci 2003; 73(24):3043-3072.
67. Drutel G, Peitsaro N, Karlstedt K et al. Identification of rat H_3 receptor isoforms with different brain expression and signaling properties. Mol Pharmacol 2001; 59(1):1-8.
68. Clark EA, Hill SJ. Sensitivity of histamine H_3 receptor agonist-stimulated [35S]GTP gamma[S] binding to pertussis toxin. Eur J Pharmacol 1996; 296(2):223-225.
69. Wieland K, Bongers G, Yamamoto Y et al. Constitutive activity of histamine $H_{(3)}$ receptors stably expressed in SK-N-MC cells: display of agonism and inverse agonism by $H_{(3)}$ antagonists. J Pharmacol Exp Ther 2001; 299(3):908-914.
70. Morisset S, Rouleau A, Ligneau X et al. High constitutive activity of native H_3 receptors regulates histamine neurons in brain. Nature 2000; 408(6814):860-864.
71. Uveges AJ, Kowal D, Zhang Y et al. The role of transmembrane helix 5 in agonist binding to the human H_3 receptor. J Pharmacol Exp Ther 2002; 301(2):451-458.
72. Cogé F, Guenin SP, Audinot V et al. Genomic organization and characterization of splice variants of the human histamine H_3 receptor. Biochem J 2001; 355(Pt 2):279-288.
73. Gomez-Ramirez J, Ortiz J, Blanco I. Presynaptic H_3 autoreceptors modulate histamine synthesis through cAMP pathway. Mol Pharmacol 2002; 61(1):239-245.
74. Sanchez-Lemus E, Arias-Montano JA. Histamine H_3 receptor activation inhibits dopamine D1 receptor-induced cAMP accumulation in rat striatal slices. Neurosci Lett 2004; 364(3):179-184.
75. Bakker RA, Lozada AF, van Marle A et al. Discovery of naturally occurring splice variants of the rat histamine H_3 receptor that act as dominant-negative isoforms. Mol Pharmacol 2006; 69(4):1194-1206.
76. Bongers G, Sallmen T, Passani MB et al. The Akt/GSK-3beta axis as a new signaling pathway of the histamine H(3) receptor. J Neurochem 2007; 103(1):248-258.
77. Thurmond RL, Gelfand EW, Dunford PJ. The role of histamine H_1 and H_4 receptors in allergic inflammation: the search for new antihistamines. Nat Rev Drug Discov 2008; 7(1):41-53.
78. Jablonowski JA, Grice CA, Chai W et al. The first potent and selective non-imidazole human histamine H_4 receptor antagonists. J Med Chem 2003; 46(19):3957-3960.
79. Liu C, Wilson SJ, Kuei C et al. Comparison of human, mouse, rat and guinea pig histamine H_4 receptors reveals substantial pharmacological species variation. J Pharmacol Exp Ther 2001; 299(1):121-130.
80. Lim HD, Smits RA, Bakker RA et al. Discovery of S-(2-guanidylethyl)-isothiourea (VUF 8430) as a potent nonimidazole histamine H_4 receptor agonist. J Med Chem 2006; 49(23):6650-6651.
81. Thurmond RL, Desai PJ, Dunford PJ et al. A potent and selective histamine H_4 receptor antagonist with anti-inflammatory properties. J Pharmacol Exp Ther Apr 2004; 309(1):404-413.
82. Coruzzi G, Adami M, Guaita E et al. Anti-inflammatory and antinociceptive effects of the selective histamine H_4-receptor antagonists JNJ7777120 and VUF6002 in a rat model of carrageenan-induced acute inflammation. Eur J Pharmacol 2007; 563(1-3):240-244.
83. Zhang M, Thurmond RL, Dunford PJ. The histamine $H_{(4)}$ receptor: a novel modulator of inflammatory and immune disorders. Pharmacol Ther 2007; 113(3):594-606.
84. de Esch IJ, Thurmond RL, Jongejan A et al. The histamine H_4 receptor as a new therapeutic target for inflammation. Trends Pharmacol Sci 2005; 26(9):462-469.
85. Coge F, Guenin SP, Rique H et al. Structure and expression of the human histamine H_4-receptor gene. Biochem Biophys Res Commun 2001; 284(2):301-309.
86. Nakamura T, Itadani H, Hidaka Y et al. Molecular cloning and characterization of a new human histamine receptor, HH_4R. Biochem Biophys Res Commun 2000; 279(2):615-620.
87. Hofstra CL, Desai PJ, Thurmond RL et al. Histamine H_4 receptor mediates chemotaxis and calcium mobilization of mast cells. J Pharmacol Exp Ther 2003; 305(3):1212-1221.
88. Nakayama T, Kato Y, Hieshima K et al. Liver-expressed chemokine/CC chemokine ligand 16 attracts eosinophils by interacting with histamine H_4 receptor. J Immunol 2004; 173(3):2078-2083.

Chapter 3

Histamine Synthesis and Lessons Learned from Histidine Decarboxylase Deficient Mice

Hiroshi Ohtsu*

Abstract

This chapter summarizes the information about the transcriptional regulation of histidine decarboxylase (HDC), which is the catabolic enzyme of histamine synthesis, and the activity of histamine in vivo as clarified using HDC gene deficient mice (HDC-KO). The research of the regulatory mechanism of histamine synthesis has been focused on transcriptional and posttranslational aspects. The generation of HDC-KO mice clarified several new pathophysiological functions of histamine. It is now recognized that the activity of histamine is not limited to allergic, peptic and neurological functions as in the old paradigm, but extends to other fields such as cardiology, immunology and infectious diseases. Therefore, this chapter will focus on these newly revealed functions of histamine. For example, histamine was known to be involved in the effector phase of allergic responses, but a role has now been shown in the sensitization phases and in innate immunity. In the allergic bronchial asthma model using HDC-KO mice it was found that histamine positively controls eosinophilia, but not bronchial hypersensitivity. The effect on eosinophils was afterwards shown to be mediated through the activity of the histamine H_4 receptor. The recent advances in the understanding of histamine synthesis and the activity of HDC have dramatically expanded our understanding of the scope of histamine function.

Introduction

Histamine, 2-(4-imidazole)-ethylamine, has been regarded as one of the most important biogenic amines since the original pharmacological studies in 1910 and 1911.[1,2] Histamine regulates smooth muscle contraction, immune responses, vascular permeability, neurotransmission and the stimulation of gastric acid secretion. Histamine is synthesized from histidine through oxidative decarboxylation by histidine-decarboxylase (HDC; EC 4.1.1.22), a pyridoxal 5′-phosphate (PLP)-dependent enzyme.[3] HDC is expressed in the liver of developing fetuses and in the stomach, brain, thymus, kidney, spleen and bones. Restricted and cell-specific expression of HDC in peripheral tissues is controlled both transcriptionally by processes such as DNA methylation[4,5] and posttranslationally where levels can be controlled by the ubiquitin-proteasome system,[6,7] caspases[8] and other mechanisms. At the transcriptional level, expression of HDC is regulated by various kinds of stimuli including gastrin,[9] phorbol esters like phorbol 12-myristate-13-acetate (PMA),[9-12] oxidative stress[13] and thrombopoietin.[14]

Histamine exerts its activity through four different G-protein-coupled receptors.[15-18] The histamine H_1 receptor (H_1R) couples to Gq G-proteins and leads to the phosphoinositol hydrolysis pathway. The H_1R is known to be responsible for the acute inflammatory responses. Antagonists

*Hiroshi Ohtsu—Tohoku University, Sendai, 980-8579, Japan.
Email: hiroshi.ohtsu@qse.tohoku.ac.jp

Histamine in Inflammation, edited by Robin L. Thurmond.

of the H_1R are used for treatment of allergic states. The H_2 receptor (H_2R) links to Gs G-proteins to activate adenylate cyclase, H_3 receptors (H_3R) couple to Gi G-proteins to inhibit the adenylate cyclase pathway and the H_4 receptor (H_4R) couples to Gi/o G-proteins and leads to increases in calcium. The representative function of H_2R pathway is gastric acid secretion. H_3R are responsible for the inhibition of neuronal histamine release. The functional role of the H_4R is an emerging field that has been energetically investigated in recent years.[19]

It has been 15 years since the cloning of HDC and much new information on its transcriptional regulation has accumulated. This chapter will compile all of this information and review the function and regulation of HDC. Moreover, this new data has also led to a better understand of the role of histamine and this too will be reviewed.

HDC Transcriptional Regulation

HDC are PLP-dependent enzymes in mammals and Gram-negative bacteria. It was deduced that the mammalian native enzyme catalyzed its activity in the form of homodimer (53-54 kDa/subunit).[20,21] However, the mammalian cDNAs were found to encode 74 kDa peptides[22] and therefore it became clear that different active HDC peptides (53-74 kDa/subunit) exist in vivo.[23,24] After cloning of HDC gene it was found that the human gene was unique per haploid genome.[25] Many studies on mammalian HDC transcriptional regulation have been devoted to characterize the mechanisms operating in gastric cells under different stimuli and circumstances (gastrin, *Helicobacter pylori* infection, oxidative stress, etc).[26] Although one may expect tissue-specific and time-specific transcriptional regulation in the HDC gene, this information is still very limited.[5]

Epigenetic Regulation of HDC Gene Expression

In hematopoietic cell lineages, HDC gene expression has been reported in mast cells and basophils. In an attempt to discover how HDC gene expression is regulated in these cells, we found that the human HDC-promoter region in HDC-expressing cell lines is selectively unmethylated.[5] A correlation between HDC expression and hypomethylation was also found in primary mast cells and methylation of a HDC promoter-luciferase reporter construct resulted in decreased luciferase activity in a transient expression system. Later on we confirmed the importance of methylation in regulating the activity of the mouse promoter.[4] HDC gene expression is strongly induced in the mouse immature mast cell line P815 after incubation in the peritoneal cavity.[27] The induction of gene expression is correlated with the demethylation of the promoter. Consistently, forced demethylation by 5-azacytidine treatment induced high expression of HDC mRNA in P815 cells. These data suggest that DNA methylation may be one of the major mechanisms that regulate HDC gene expression and may be of functional importance in the development of mast cells.

In addition to DNA methylation, there is also evidence that histone acetyltransferases (HATs) and histone deacetylases (HDACs) regulate HDC gene expression. After the demonstration that KLF4 represses the HDC gene promoter,[26] Tip60 was identified as a KLF4-interacting protein. Transfection experiments suggest that Tip60 inhibits HDC-promoter activity. The data also suggests that Tip60 functions as a corepressor of KLF4 in the regulation of HDC promoter activity.[28]

Histamine production in enterochromaffin-like cells in the stomach is controlled by gastrin[29] and, in turn, histamine controls gastric acid secretion by activating the proton pump in parietal cells through H_2R activation.[30] This is the rationale for the use of H_2R antagonists for treating peptic ulcer disease. Studies in HDC-KO mice that are therefore histamine-deficient have confirmed that de novo histamine synthesis is essential for gastric acid secretion induced by gastrin, but does not play a role in vagal release of acetylcholine that also participates in acid production.[31] Since gastrin treatment decreased the association of KLF4, Tip60 and HDAC7 with HDC promoter, the decrease of repressive complexes formed by these factors might suppress HDC transcription.[28] In this report a new technology, chromatin immunoprecipitation (ChIP), was used that allows the in vivo identification of direct transcription factor-binding sites in the context of chromatin and, therefore, avoids many of the previous problems. In theory, ChIP could be used to investigate any target on chromatin against which an antibody can be raised and, consequently, it has successfully been used to identify regions of the genome associated with specific transcription factors, cofactors, histone modifications and DNA methylation. By using ChIP analysis it was found that transcription factors bind the promoter region of HDC and not to the first or second intron. The binding was disrupted by gastrin treatment suggesting that Tip60 and HDAC7 act as corepressors of KLF4 in the regulation of HDC gene expression.[28] It has also been reported that the promoter of the HDC gene was activated by hypoxia inducible factor-1 in a human mast cell line and this activation seems to occur during hypoxic stress.[32]

L-Histidine Decarboxylase Gene Knockout Mice

HDC-KO mice lack the ability to synthesize histamine and, therefore, cannot undergo histamine-dependent activation through any of the four known histamine receptors.[33] These mice have been used to clarify the role of histamine in various conditions including anaphylactic responses and allergic inflammation,[34-36] as well as several neurophysiologic functions.[37,38] Because it is difficult to achieve complete and long-lasting elimination of the effects of histamine in vivo by pharmacological approaches, HDC-KO mice provide an excellent tool for studying the effect of chronic deprivation of histamine in disease models.

Histamine and Immunity

Histamine is well known for its role in immunological reactions. The development of H_1R and H_2R knockout mice has clarified the role of histamine in T cells.[39,40] Histamine has been proved to be important as a regulator in Th1 and Th2 cells.[39] Gutzmer et al. have reported that human dendritic cells express all four histamine receptors (H_1R to H_4R). H_2R and H_4R stimulation suppresses IL-12p70 production and mediates chemotaxis in human dendritic cells.[41]

It is also clear that histamine functions in innate immunity since it mediates NK cell chemotaxis[42] and cytokine release from invariant natural killer T cells (iNKT) through the H_4R.[43] Histamine-free HDC-KO mice have a numerical and functional deficit in iNKT cells as evidenced by a drastic decrease in IL-4 and IFN-γ production by these cells. This deficiency was established both by measuring cytokine levels in the serum and intracellularly in gated iNKT cells. The defect was due to the lack of histamine since a single injection of histamine into HDC-KO mice sufficed to restore normal IL-4 and IFN-γ production. Histamine-induced functional recovery was mediated mainly through the H_4R since it could be abrogated by a selective H_4R antagonist and by the demonstration of a similar iNKT cell deficit in H_4R-deficient (H_4R-KO) mice. These findings identify a novel function of histamine through the H_4R in modulating iNKT cell functions and, therefore, may contribute the initial host defense mechanism. Furthermore since the production of IL-4 is important for antibody class switching to produce IgE, it is possible that histamine is not only a canonical allergic effecter molecule, but can also regulate the afferent phase of the allergic state as well.

Histamine in Wound Healing

The absence of histamine in HDC-KO mice resulted in delayed cutaneous wound healing and exogenously administered histamine was able to restore this response.[44] Furthermore, overproduction of histamine in HDC gene-transgenic mice lead to accelerate healing compared to wild-type mice (Fig. 1). These results indicate that histamine accelerates cutaneous wound healing. Macrophage recruitment and angiogenesis at the wound edge were specifically impaired in HDC-KO mice and histamine treated wounds in HDC-KO mice demonstrated increased macrophage recruitment and angiogenesis. Since macrophages are an important cellular component for the wound healing process, it was postulated that histamine, whether directly or indirectly, accelerates the recruitment of macrophage to the wound. The protein levels of basic fibroblast growth factor (bFGF) at the wound edge was higher in wild-type mice, especially on the 3rd and 5th day of wound healing, compared to HDC-KO mice. Topically administered SU5402, a specific antagonist to fibroblast growth factor receptor-1 (FGFR1) tyrosine kinase, to the wound surface suppressed the wound healing in wild-type mice, but not in HDC-KO mice. Moreover, SU5402 reduced macrophage recruitment and angiogenesis in wild-type mice. From these observations it was concluded that the accelerated

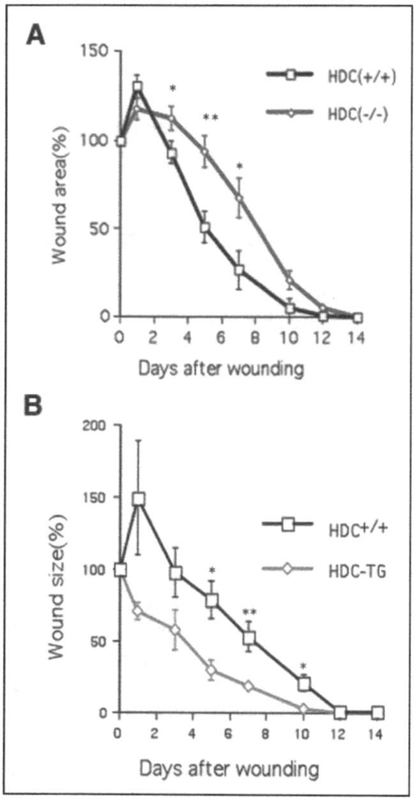

Figure 1. Histamine affects on the wound healing process. A) Wound areas in HDC-KO (HDC+/+) and wild-type (HDC–/–) mice were determined by tracing the wound margin using a transparent sheet. Results are means ± SEM, $n = 8$, for each time point and group. *$P < 0.05$, **$P < 0.01$ compared with wild-type mice. B) Wound area of wild-type and HDC transgenic (HDC-Tg) mice. Results are means ± SEM, $n = 6$ for the wild-type, $n = 5$ for the HDC-Tg mice. *$P < 0.05$, **$P < 0.01$ compared with wild-type mice. The figure is modified from reference 44.

wound healing activity of histamine was mediated by the activity of bFGF, which leads to angiogenesis and macrophage recruitment in the wound healing process.

Histamine in Malaria

There are several lines of evidence suggesting that the existence of histamine is beneficial for parasites to spread through the vasculature, perhaps due to its vasodilatory effects.[45] Furthermore, histamine can increase endothelial expression of thrombomodulin through the H_1R.[46] Thrombomodulin is important for the sequestration of the parasitized erythrocyte through its binding to anticoagulants. From the inoculation of Plasmodium sporozoites via Anopheles mosquito bites to the development of blood-stage parasites, a hallmark of the host response is an inflammatory reaction characterized by elevated histamine levels in the serum and tissues. After consideration of the proinflammatory and immunosuppressive activities associated with histamine, it can be postulated that histamine participates in malaria pathogenesis. Combined genetic and pharmacological approaches demonstrated that histamine binding to H_1R and H_2R, but not H_3R and H_4R increases the susceptibility of mice to infection with Plasmodium.[47] To further understand the role of histamine in malaria pathogenesis, HDC-KO mice were used. HDC-KO mice were highly resistant to severe malaria whether infected by mosquito bites or via injection of infected erythrocytes.[47] HDC-KO mice displayed resistance to two lethal strains: Plasmodium berghei (Pb) ANKA, which triggers cerebral malaria (CM) and Pb NK65, which causes death without neurological symptoms. The resistance of HDC-KO mice to CM was associated with a preserved blood-brain barrier integrity. Also the resistance seemed to be associated with the absence of infected erythrocyte aggregation in the brain vessels. No infiltration of CD4 and CD8 T cells was observed in the brains of infected HDC-KO mice, although this was observed in wild-type mice after infection. These results suggest that histamine-mediated signaling contributes to malaria pathogenesis and that understanding the basis for these biological effects may lead to novel therapeutic strategies to alleviate the severity of malaria.

Histamine in Crohn's Disease

Crohn's disease (CD), one of the two major types of inflammatory bowel disease, is a chronic inflammatory disorder of unknown etiology that affects any part of the gastrointestinal tract from the mouth to anus, however, the small intestine and colon are most commonly involved. It has been considered that CD is a multifactoral disease with genetic background, environmental factors and immunological responses all contributing.

Although historically CD is rare in Asian countries, the number of patients with CD has increased during the past decade. One of the reasons for the change is thought to be due to changes in diet and exposure to new dietary antigens that may stimulate the mucosal immune response. Patients with active CD respond to bowel rest along with total parenteral nutrition (TPN). In addition, enteral nutrition in the form of elemental or peptide-based preparations is as effective as glucocorticoids or TPN. Although the therapeutic mechanisms of elemental diets (ED) remain unclear, some possibilities have been suggested: (1) a low-antigenic diet reduces the mucosal immune response; (2) a low-fat diet is less pro-inflammatory; and (3) that ED alters the enteric flora population.

Since ED may have mechanisms to suppress intestinal inflammation directly, the effect of amino acids that are the main component of ED was assessed. Histidine reduced histologic damage in a model of CD.[48] Furthermore, histidine, but not lysine or alanine, inhibited the production of TNFα and IL-6 by LPS-stimulated peritoneal macrophages in a concentration-dependent manner. Since intracellular histidine is metabolized to histamine by HDC, we hypothesized that histamine might be the substance which ameliorates CD in histidine supplemented ED. However, histidine was still able to reduce production of TNFα and IL-6 in LPS-stimulated HDC-deficient peritoneal macrophages.[48] Thus, the anti-inflammatory effect of histidine seemed to be independent of histamine synthesis.

Histamine in Allergic Bronchial Asthma

Histamine is a major mediator that elicits a number of the acute physiologic responses in allergic asthma.[49] Its role in asthma is supported by several pieces of evidence including the release of histamine from cells participating in allergic responses, the reproduction of the features of allergic inflammation by application of histamine, the reduction of allergic inflammation by histamine receptor antagonists and the reduced eosinophilia in mice genetically modified not to synthesize histamine.[35] Mast cells, basophils, enterochromaffin-like cells in gastric wall and neurons in tuberomammillary nucleus in hypothalamus are major sources of histamine at the cellular level. Mast cells and basophils are the postulated major sources of histamine in allergic reaction. Histamine release from these cells is triggered by the interaction of an allergen with specific immunoglobulin E (IgE) bound to the high-affinity IgE receptor on the cell membrane or by nonspecific stimuli including exercise or cold, dry air. The actions of histamine in allergic asthma may be rather complicated since it not only has the direct actions on smooth muscle and sensory nerves, but also can have indirect activity on vagal reflexes that cause cough.

The role of histamine in immunology has been a major topic of study over the past few years. Previous studies have suggested that histamine enhances Th2 responses through modulation of dendritic cell function and regulation of IL-10 and IL-12 production. Dendritic cells express H_1R, H_2R and H_4R[50] and their exposure to histamine induces a shift toward the DC2 phenotype with regard to the expression of cytokines and chemokines that promote Th2 immune responses.[51,52]

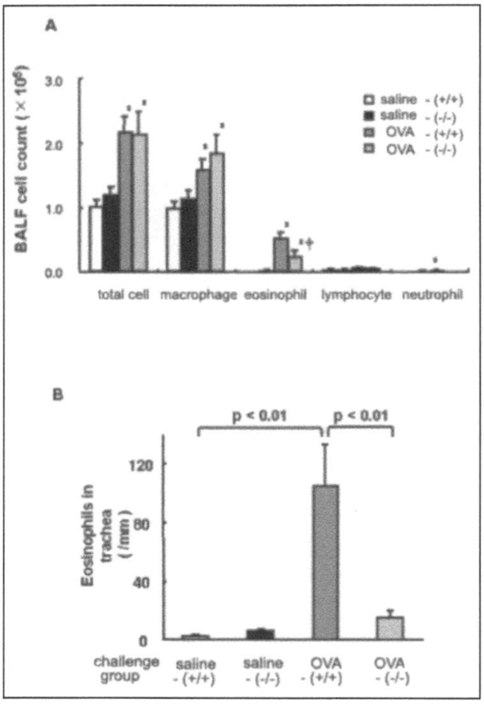

Figure 2. A) Total cell and differential leukocyte counts in the bronchoalveolar lavage fluid BALF in a mouse allergic asthma model. On the 3rd day after the ovalbumin (OVA) inhalation challenge, the mice were anesthetized and the recovered fluid from bronchial tract was analyzed. Each value indicates mean ± SEM in 1 ml BALF of 8 to 11 mice (Mann-Whitney U test). HDC-KO are designated as (–/–) and wild-type mice as (+/+). B) Eosinophil counts in the submucosal area of the trachea. Eosinophil number within the submucosal area encircling the trachea was counted and expressed as number/mm basement membrane. Each value indicates mean ± SEM of five to six animals (Mann-Shitney U test). Modified with permission from reference 35, ©2003 American Thoracic Society.

Histamine inhibits IL-12 production, while enhancing IL-10 synthesis in lipopolysaccharide-treated leukocytes.[53,54] These observations suggest a Th2-promoting influence of histamine. In contrast, studies with mice bearing a targeted deletion of the H_1R show reduced production of IFNγ and increased IL-4 and IL-13 secretion, results more consistent with the Th1-polarizing function of this receptor.[39] Recently, the H_4R was shown to play an important role in allergic lung inflammation with effects on Th2 responses and the recruitment of lung eosinophils and lymphocytes.[55] Blockade of the H_4R on dendritic cells leads to decreases in cytokine and chemokine production and limits their ability to induce Th2 responses in T cells. In addition effects on innate immune responses such as those on iNKT cells already mentioned indicate that histamine plays important role not only during the efferent phase but also for the afferent phase of allergic responses.

We previously examined the role of endogenous histamine on eosinophilic recruitment and hyperresponsiveness in allergic bronchial asthma mouse model using HDC-KO mice.[35] Histamine levels in the airways in HDC-KO mice were largely diminished compared with wild-type mice. Inhalation challenge with ovalbumin (OVA) in OVA-sensitized wild-type mice caused total cell, macrophage and eosinophil accumulation in the lung as well as airway hyper-responsiveness to methacholine three days after the challenge. The eosinophil recruitment to the lung was significantly reduced in HDC-KO mice compared with wild-type mice(Fig. 2).

In the bone marrow proliferation of eosinophils was induced after OVA challenge in wild-type mice; however, the proliferation was significantly suppressed in HDC-KO mice. In contrast, airway hyper-responsiveness was not suppressed in HDC-KO mice. These results suggest that endogenous histamine is involved in the accumulation of eosinophils into the airways after allergic challenge, possibly via effects in the bone marrow. Since histamine has eosinophil chemotactic activity via H_4R,[56] reduced eosinophilia in HDC-KO mice could be explained through the activity via H_4R. However, allergen-induced airway hyper-responsiveness occurred independently of airway eosinophilia in this model.

Histamine in Systemic Anaphylaxis Model

Compared with other allergic disorders, such as asthma, allergic dermatitis and allergic rhinitis, the pathophysiology of anaphylaxis seems relatively simple. Antigen cross-links antibody molecules, activating immunoglobulin receptors on inflammatory cells and causing them to release mediators that increase vascular permeability and cause smooth muscle contractions that produce urticaria, hypotension, dypnea, abdominal cramping and diarrhea.[57] Mouse systemic anaphylaxis reaction is IgE-dependent and manifested by hypotension, airway obstruction and hypothermia.[36] The Fcε receptor is expressed on the plasma membrane of mast cells and is bound to IgE antibodies.[58-62] Antigen binding to IgE molecules leads to cross-linking of the Fcε receptor, which in turn leads to the secretion of the intragranular content into the environmental fluid, leading to the symptoms observed in systemic anaphylaxis. The reaction starts from a few minutes to 30 minutes after the allergen challenge and it leads to an increase in vascular permeability, contraction of smooth muscle and an increase in mucin secretion. The final results are circulatory collapse, which is observed as hypotension, an increase in heart rate and a decrease in peripheral blood resistance, sometimes leading to death.

To elucidate the role of histamine in each of these symptoms, we induced a passive systemic anaphylaxis reaction in HDC-KO mice and wild-type mice and compared the changes in body temperature, blood pressure and respiratory function.[36] Blood pressure dropped in both HDC-KO and wild-type mice. However, decreases in respiratory frequency, body temperature and elongation of expiratory respiration time occurred only in wild-type mice (Fig. 3). Therefore, in this model of passive systemic anaphylaxis reaction via the Fcε receptor, respiratory frequency, expiratory time and body temperature were shown to be controlled by the activity of histamine, but its contribution to blood pressure was small. Since mast cell deficient W/Wv mice did not show any decrease in body temperature compared to control mice, it can be concluded that histamine derived from mast cells contributes to the change of body temperature.

Histamine in Atherosclerosis

HDC mRNA levels increase during the progression of atherosclerosis in the human aorta. HDC protein localized to macrophage-derived foam cells and mononuclear cells including lymphocytes.[63] The effect of histamine depends on the vascular size and localization. In capillaries, histamine distends the vessel wall and exerts an inflammatory reaction such as extravasation of blood. In contrast the muscular arteries, including coronary and mesenteric arteries, are constricted by histamine, owing to the contraction of medial smooth muscle cell (SMC). Histamine also has effects on the proliferation of SMC.[64] To clarify the role of histamine-producing cells and its origin in atherosclerosis, we investigated HDC expression in atherosclerotic arteries after the mouse had received a bone marrow transfusion from green fluorescent protein (GFP)-transgenic mice.[65] Two different atherosclerosis models were investigated—a ligation-induced vascular injury model and a cuff-induced vascular injury model. In the ligation model, the neointima of atherosclerotic carotid arteries contained bone marrow-derived HDC+/+ cells expressing macrophage or smooth muscle antigens. In contrast in the cuff replacement model, the HDC+/+ bone marrow-derived cells, which were positive for Mac-3, were mainly located in the adventitia. After inducing atherosclerotic lesions, HDC-KO mice showed reduced neointimal thickening and a decreased intima-to-media ratio in comparison to wild-type mice. These results indicate that histamine produced from bone

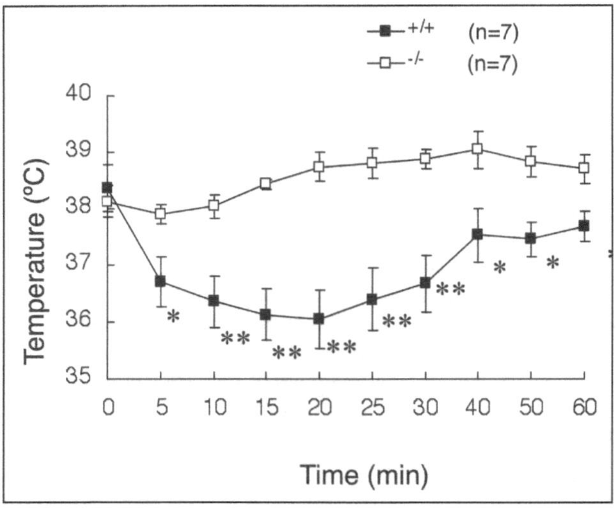

Figure 3. Change of body temperature in HDC-KO (–/–) and wild-type (+/+) mice after the passive systemic anaphylaxis reaction. Body temperature was monitored using thermometer probe situated in anal. *P < 0.05 and **P < 0.01 in comparison with HDC-KO mice. Reproduced with permission from reference 36, ©2002 from Elsevier.

marrow-derived progenitor cells, which could transdifferentiate into SMC- or macrophage-like cells, are important for the formation of neointima and atheromatous plaques.

Histamine Uptake into and Release from Histamine Producing Cells

A number of cells have been shown to take up histamine. For example a medullary population of basophils containing few granules have been shown to take up histamine from the environment.[66] Mast cells from HDC-KO mice were also demonstrated to store exogenous histamine immediately after incubation with histamine and release the stored histamine after anti-DNP IgE and DNP-HSA treatment.[67]

The uptake of the histamine into the basophil/mast cell granules involves transport through two distinct membranes—the plasma membrane and the vesicular membrane. It was shown that the bidirectional organic cation transporter 3 (OCT3/Slc22a3) has a role in histamine uptake.[68] In addition the vesicular monoamine transporter 2 (VMAT2) is involved in the vesicular membrane transport system, since mast cells prepared from VMAT2-deficient mice showed reduced release of histamine.[69] The vesicular membrane transport system appeared to be more efficient than plasma membrane transport, because the amount of histamine in the granular fraction was much higher than in cytosolic fraction. Therefore, once histamine has been absorbed by the mast cell, it is transported into the secretory granules quite rapidly.[67] OCT3 also participates in the control of basophil function since exogenous histamine inhibits not only its own synthesis, but also that of IL-4, IL-6 and IL-13.[68]

Conclusion

This chapter has summarized the recently clarified mechanisms of histidine decarboxylase regulation and the insights gained using histidine decarboxylase gene knockout mice on the varied roles of histamine in vivo. Since in vivo the synthesis of histamine from histidine is catalyzed solely by histidine decarboxylase, the amount of histamine is suppressed closely to null levels by knocking out the gene and hence the HDC-KO have been a useful tool for observing altered phenotypes in the absence of histamine. The availability of the gene-manipulated mice along with the aid of selective pharmacological agents have been powerful tools for uncovering and clarifying the various phenomena mediated by histamine in vivo. The information gained by using such approaches can provide basic knowledge that can be used for exploring the direction of clinical studies.

References

1. Dale HH, Laidlaw PP. The physiological action of beta-iminazolylethylamine. J Physiol 1910; 41(5):318-344.
2. Dale HH, Laidlaw PP. Further observations on the action of beta-iminazolylethylamine. J Physiol 1911; 43(2):182-195.
3. Moya-Garcia AA, Medina MA, Sanchez-Jimenez F. Mammalian histidine decarboxylase: from structure to function. Bioessays 2005; 27(1):57-63.
4. Suzuki-Ishigaki S, Numayama-Tsuruta K, Kuramasu A et al. The mouse L-histidine decarboxylase gene: structure and transcriptional regulation by CpG methylation in the promoter region. Nucleic Acids Res 2000; 28(14):2627-2633.
5. Kuramasu A, Saito H, Suzuki S et al. Mast cell-/basophil-specific transcriptional regulation of human L-histidine decarboxylase gene by CpG methylation in the promoter region. J Biol Chem 1998; 273(47):31607-31614.
6. Fleming JV, Fajardo I, Langlois MR et al. The C-terminus of rat L-histidine decarboxylase specifically inhibits enzymic activity and disrupts pyridoxal phosphate-dependent interactions with L-histidine substrate analogues. Biochem J 2004; 381(Pt 3):769-778.
7. Olmo MT, Urdiales JL, Pegg AE et al. In vitro study of proteolytic degradation of rat histidine decarboxylase. Eur J Biochem 2000; 267(5):1527-1531.
8. Furuta K, Nakayama K, Sugimoto Y et al. Activation of histidine decarboxylase through posttranslational cleavage by caspase-9 in a mouse mastocytoma P-815. J Biol Chem 2007; 282(18):13438-13446.
9. Zhang Z, Hocker M, Koh TJ et al. The human histidine decarboxylase promoter is regulated by gastrin and phorbol 12-myristate 13-acetate through a downstream cis-acting element. J Biol Chem 1996; 271(24):14188-14197.
10. Hocker M, Henihan RJ, Rosewicz S et al. Gastrin and phorbol 12-myristate 13-acetate regulate the human histidine decarboxylase promoter through Raf-dependent activation of extracellular signal-regulated kinase-related signaling pathways in gastric cancer cells. J Biol Chem 1997; 272(43):27015-27024.
11. Hocker M, Zhang Z, Fenstermacher DA et al. Rat histidine decarboxylase promoter is regulated by gastrin through a protein kinase C pathway. Am J Physiol 1996; 270(4 Pt 1):G619-633.
12. Ohgoh M, Yamamoto J, Kawata M et al. Enhanced expression of the mouse L-histidine decarboxylase gene with a combination of dexamethasone and 12-O-tetradecanoylphorbol-13-acetate. Biochem Biophys Res Commun 1993; 196(3):1113-1119.
13. Hocker M, Rosenberg I, Xavier R et al. Oxidative stress activates the human histidine decarboxylase promoter in AGS gastric cancer cells. J Biol Chem 1998; 273(36):23046-23054.
14. Pacilio M, Debili N, Arnould A et al. Thrombopoietin induces histidine decarboxylase gene expression in c-mpl transfected UT7 cells. Biochem Biophys Res Commun 2001; 285(5):1095-1101.
15. Fukui H, Fujimoto K, Mizuguchi H et al. Molecular cloning of the human histamine H1 receptor gene. Biochem Biophys Res Commun 1994; 201(2):894-901.
16. Gantz I, Munzert G, Tashiro T et al. Molecular cloning of the human histamine H2 receptor. Biochem Biophys Res Commun 1991; 178(3):1386-1392.
17. Lovenberg TW, Roland BL, Wilson SJ et al. Cloning and functional expression of the human histamine H3 receptor. Mol Pharmacol 1999; 55(6):1101-1107.
18. Oda T, Morikawa N, Saito Y et al. Molecular cloning and characterization of a novel type of histamine receptor preferentially expressed in leukocytes. J Biol Chem 2000; 275(47):36781-36786.
19. Huang JF, Thurmond RL. The new biology of histamine receptors. Curr Allergy Asthma Rep 2008; 8(1):21-27.
20. Taguchi Y, Watanabe T, Kubota H et al. Purification of histidine decarboxylase from the liver of fetal rats and its immunochemical and immunohistochemical characterization. J Biol Chem 1984; 259(8):5214-5221.
21. Martin SA, Bishop JO. Purification and characterization of histidine decarboxylase from mouse kidney. Biochem J 1986; 234(2):349-354.
22. Joseph DR, Sullivan PM, Wang YM et al. Characterization and expression of the complementary DNA encoding rat histidine decarboxylase. Proc Natl Acad Sci USA 1990; 87(2):733-737.
23. Yatsunami K, Tsuchikawa M, Kamada M et al. Comparative studies of human recombinant 74- and 54-kDa L-histidine decarboxylases. J Biol Chem 1995; 270(51):30813-30817.
24. Fleming JV, Wang TC. The production of 53-55-kDa isoforms is not required for rat L-histidine decarboxylase activity. J Biol Chem 2003; 278(1):686-694.

25. Yatsunami K, Ohtsu H, Tsuchikawa M et al. Structure of the L-histidine decarboxylase gene. J Biol Chem 1994; 269(2):1554-1559.
26. Ai W, Liu Y, Langlois M et al. Kruppel-like factor 4 (KLF4) represses histidine decarboxylase gene expression through an upstream Sp1 site and downstream gastrin responsive elements. J Biol Chem 2004; 279(10):8684-8693.
27. Ohtsu H, Kuramasu A, Suzuki S et al. Histidine decarboxylase expression in mouse mast cell line P815 is induced by mouse peritoneal cavity incubation. J Biol Chem 1996; 271(45):28439-28444.
28. Ai W, Zheng H, Yang X et al. Tip60 functions as a potential corepressor of KLF4 in regulation of HDC promoter activity. Nucleic Acids Res. 2007; 35(18):6137-6149.
29. Norlen P, Ericsson P, Kitano M et al. The vagus regulates histamine mobilization from rat stomach ECL cells by controlling their sensitivity to gastrin. J Physiol 2005; 564(Pt 3):895-905.
30. Prinz C, Zanner R, Gratzl M. Physiology of gastric enterochromaffin-like cells. Annu Rev Physiol 2003; 65:371-382.
31. Tanaka S, Hamada K, Yamada N et al. Gastric acid secretion in L-histidine decarboxylase-deficient mice. Gastroenterology 2002; 122(1):145-155.
32. Jeong HJ, Moon PD, Kim SJ et al. Activation of hypoxia-inducible factor-1 regulates human histidine decarboxylase expression. Cell Mol Life Sci 2009; 66(7):1309-1319.
33. Ohtsu H, Tanaka S, Terui T et al. Mice lacking histidine decarboxylase exhibit abnormal mast cells. FEBS Lett 2001; 502(1-2):53-56.
34. Hirasawa N, Ohtsu H, Watanabe T et al. Enhancement of neutrophil infiltration in histidine decarboxylase-deficient mice. Immunology 2002; 107(2):217-221.
35. Koarai A, Ichinose M, Ishigaki-Suzuki S et al. Disruption of L-histidine decarboxylase reduces airway eosinophilia but not hyperresponsiveness. Am J Respir Crit Care Med 2003; 167(5):758-763.
36. Makabe-Kobayashi Y, Hori Y, Adachi T et al. The control effect of histamine on body temperature and respiratory function in IgE-dependent systemic anaphylaxis. J Allergy Clin Immunol 2002; 110(2):298-303.
37. Parmentier R, Ohtsu H, Djebbara-Hannas Z et al. Anatomical, physiological and pharmacological characteristics of histidine decarboxylase knock-out mice: evidence for the role of brain histamine in behavioral and sleep-wake control. J Neurosci 2002; 22(17):7695-7711.
38. Dere E, De Souza-Silva MA, Spieler RE et al. Changes in motoric, exploratory and emotional behaviours and neuronal acetylcholine content and 5-HT turnover in histidine decarboxylase-KO mice. Eur J Neurosci 2004; 20(4):1051-1058.
39. Jutel M, Watanabe T, Klunker S et al. Histamine regulates T-cell and antibody responses by differential expression of H1 and H2 receptors. Nature 2001; 413(6854):420-425.
40. Pedotti R, De Voss JJ, Steinman L et al. Involvement of both 'allergic' and 'autoimmune' mechanisms in EAE, MS and other autoimmune diseases. Trends Immunol 2003; 24(9):479-484.
41. Gutzmer R, Diestel C, Mommert S et al. Histamine H4 receptor stimulation suppresses IL-12p70 production and mediates chemotaxis in human monocyte-derived dendritic cells. J Immunol 2005; 174(9):5224-5232.
42. Damaj BB, Becerra CB, Esber HJ et al. Functional expression of H4 histamine receptor in human natural killer cells, monocytes and dendritic cells. J Immunol 2007; 179(11):7907-7915.
43. Leite-de-Moraes MC, Diem S, Michel ML et al. Cutting edge: Histamine receptor H4 activation positively regulates in vivo IL-4 and IFN-gamma production by invariant NKT cells. J Immunol 2009; 182(3):1233-1236.
44. Numata Y, Terui T, Okuyama R et al. The accelerating effect of histamine on the cutaneous wound-healing process through the action of basic fibroblast growth factor. J Invest Dermatol 2006; 126(6):1403-1409.
45. Willadsen P, Wood GM, Riding GA. The relation between skin histamine concentration, histamine sensitivity and the resistance of cattle to the tick, Boophilus microplus. Z Parasitenkd 1979; 59(1):87-93.
46. Hirokawa K, Aoki N. Up-regulation of thrombomodulin by activation of histamine H1-receptors in human umbilical-vein endothelial cells in vitro. Biochem J 1991; 276(Pt 3):739-743.
47. Beghdadi W, Porcherie A, Schneider BS et al. Inhibition of histamine-mediated signaling confers significant protection against severe malaria in mouse models of disease. J Exp Med 2008; 205(2):395-408.
48. Andou A, Hisamatsu T, Okamoto S et al. Dietary histidine ameliorates murine colitis by inhibition of proinflammatory cytokine production from macrophages. Gastroenterology 2009; 136(2):564-574 e562.
49. White MV. The role of histamine in allergic diseases. J Allergy Clin Immunol 1990; 86(4 Pt 2):599-605.
50. Gutzmer R, Langer K, Lisewski M et al. Expression and function of histamine receptors 1 and 2 on human monocyte-derived dendritic cells. J Allergy Clin Immunol 2002; 109(3):524-531.
51. Caron G, Delneste Y, Roelandts E et al. Histamine polarizes human dendritic cells into Th2 cell-promoting effector dendritic cells. J Immunol 2001; 167(7):3682-3686.

52. Mazzoni A, Young HA, Spitzer JH et al. Histamine regulates cytokine production in maturing dendritic cells, resulting in altered T-cell polarization. J Clin Invest 2001; 108(12):1865-1873.
53. Elenkov IJ, Webster E, Papanicolaou DA et al. Histamine potently suppresses human IL-12 and stimulates IL-10 production via H2 receptors. J Immunol 1998; 161(5):2586-2593.
54. van der Pouw Kraan TC, Snijders A, Boeije LC et al. Histamine inhibits the production of interleukin-12 through interaction with H2 receptors. J Clin Invest 1998; 102(10):1866-1873.
55. Dunford PJ, O'Donnell N, Riley JP et al. The histamine H4 receptor mediates allergic airway inflammation by regulating the activation of CD4+ T-cells. J Immunol 2006; 176(11):7062-7070.
56. O'Reilly M, Alpert R, Jenkinson S et al. Identification of a histamine H4 receptor on human eosinophils—role in eosinophil chemotaxis. J Recept Signal Transduct Res 2002; 22(1-4):431-448.
57. Siraganian RP. Biochemical events in basophil/mast cell activation and medicator release. In: Adkinson NFJea, ed. Middleton's Allergy. Philadelphia: Mosby; 2003:243-276.
58. Ishizaka T, Ishizaka K. Activation of mast cells for mediator release through IgE receptors. Prog Allergy 1984; 34:188-235.
59. Ott VL, Cambier JC. Activating and inhibitory signaling in mast cells: new opportunities for therapeutic intervention? J Allergy Clin Immunol 2000; 106(3):429-440.
60. Kinet JP. The high-affinity receptor for IgE. Curr Opin Immunol 1989; 2(4):499-505.
61. Kaneko M, Schimming A, Gleich GJ et al. Ligation of IgE receptors causes an anaphylactic response and neutrophil infiltration but does not induce eosinophilic inflammation in mice. J Allergy Clin Immunol 2000; 105(6 Pt 1):1202-1210.
62. Nagai H, Abe T, Yamaguchi I et al. Role of mast cells in the onset of IgE-mediated late-phase cutaneous response in mice. J Allergy Clin Immunol 2000; 106(1 Pt 2):S91-98.
63. Higuchi S, Tanimoto A, Arima N et al. Effects of histamine and interleukin-4 synthesized in arterial intima on phagocytosis by monocytes/macrophages in relation to atherosclerosis. FEBS Lett 2001; 505(2):217-222.
64. Miyazawa N, Watanabe S, Matsuda A et al. Role of histamine H1 and H2 receptor antagonists in the prevention of intimal thickening. Eur J Pharmacol 1998; 362(1):53-59.
65. Sasaguri Y, Wang KY, Tanimoto A et al. Role of histamine produced by bone marrow-derived vascular cells in pathogenesis of atherosclerosis. Circ Res 2005; 96(9):974-981.
66. Corbel S, Schneider E, Lemoine FM et al. Murine hematopoietic progenitors are capable of both histamine synthesis and uptake. Blood 1995; 86(2):531-539.
67. Ohtsu H, Kuramasu A, Tanaka S et al. Plasma extravasation induced by dietary supplemented histamine in histamine-free mice. Eur J Immunol 2002; 32(6):1698-1708.
68. Schneider E, Machavoine F, Pleau JM et al. Organic cation transporter 3 modulates murine basophil functions by controlling intracellular histamine levels. J Exp Med 2005; 202(3):387-393.
69. Travis ER, Wang YM, Michael DJ et al. Differential quantal release of histamine and 5-hydroxytryptamine from mast cells of vesicular monoamine transporter 2 knockout mice. Proc Natl Acad Sci USA 2000; 97(1):162-167.

Chapter 4

Histamine in Allergic Rhinitis

Thomas Taylor-Clark*

Abstract

Histamine plays a major role in allergic rhinitis. In susceptible individuals, allergen induces nasal mast cell degranulation and the release of histamine into the nasal mucosa. Histamine has been detected after controlled challenges with allergen and, when administered into the nasal cavity, elicits signs and symptoms similar to those elicited by allergen. All four histamine receptors have been demonstrated in the nasal mucosa. The role of the four histamine receptors in the pathophysiology of allergic rhinitis are discussed.

Introduction

Allergic rhinitis is the result of abnormal inflammation of the nasal mucosa, characterized by nasal congestion, nasal itchiness (pruritus), sneezing and a runny nose (rhinorrhea).[1] In man, these symptoms are due to the release, into the nasal mucosa, of pro-inflammatory mediators from mast cells (and basophils) after IgE-dependent activation. The presence on the surface of mast cells of specific IgE against specific aeroallergens determines the sensitivity of individuals to a wide range of innocuous substances. These aeroallergens include substances that vary throughout the year (e.g., grass pollen—"Seasonal") and those that do not (e.g., animal dander—"Perennial") and according to the World Allergy Organization approximately 24-28% of the population of the United States and the United Kingdom suffer from some form of allergic rhinitis.

The signs and symptoms of seasonal allergic rhinitis (SAR) can be recreated in a laboratory setting, with nasal challenges of solubilized pollen sprays eliciting a two-phase response in sensitized subjects (early phase starts immediately and typically lasting up to an hour and the late phase occurs approximately 4-12 hours post challenge). Concurrent with the symptoms of congestion, pruritus, sneezing and rhinorrhea is the release of mast cell mediators such as histamine, leukotrienes, prostaglandins and tryptase. The inflammatory environment is increased by the influx of lymphocytes and granulocytes and the production of a wide array of cytokines, chemokines, growth factors and adhesion molecules. Allergic rhinitis is also characterized by airway hyperreactivity, in that nasal challenge of vasoactive or neuroactive stimuli elicit greater responses in rhinitics. In addition, a large percentage of individuals with allergic asthma also have allergic rhinitis and there is evidence that alleviation of nasal symptoms has beneficial effects on lower airway function.[2,3]

This book chapter will concentrate, where possible, on data derived from studies on human subjects and human tissue. In addition, the focus of the chapter will be on the role of histamine in seasonal allergic rhinitis. Although allergic reactions generally follow the same inflammatory pathways independent on the type of allergen, it is well established that some perennial allergens such as house dust mite have non-IgE-mediated activity which can alter the pathophysiology of the disease.

*Thomas Taylor-Clark—Department of Molecular Pharmacology and Physiology, School of Basic Biomedical Sciences, University of South Florida, 12901 Bruce B. Downs Blvd, Tampa, Florida 33612, USA. Email: ttaylorc@health.usf.edu

Histamine in Inflammation, edited by Robin L. Thurmond.

Histamine and the Early Phase Response in Allergic Rhinitis

Despite the complex inflammatory environment produced following IgE-dependent mast cell activation, histamine plays a critical role in the symptoms of allergic rhinitis. Histamine, synthesized from L-histidine by histidine decarboxylase, is present in nasal mast cells and circulating basophils and is released following allergen challenge. Histidine decarboxylase is also expressed by other immune cells, such as dendritic cells and lymphocytes.

When administered to the nasal mucosa histamine mimics the early phase of allergen challenge, causing nasal blockage, pruritus, sneezing and rhinorrhea.[4-6] These symptoms develop rapidly and fade within 15-20 minutes due to metabolism and venous clearance.

To date, four histamine receptors have been described and all four metabotropic GPCRs (H_1R, H_2R, H_3R and H_4R) have been detected in the nasal mucosa.[7] Antagonists at H_1R and H_2R have been widely used clinically to study the roles of these receptors in histamine- and allergen-induced nasal symptoms. Recently, ligands at H_3R and H_4R have been approved for some human studies. Histamine research has also benefitted from the development of knockout mouse strains of HDC as well as all four histamine receptors.

H_1 Receptors: Sensory Nerve Activation and Central Reflexes

The selective H_1R agonist betahistine, but not the selective H_2R agonist dimaprit or the H_3R agonist R-α-methylhistamine, causes sneezing, pruritus and rhinorrhea in humans.[8] The crucial role of H_1R in histamine-induced neuronal symptoms is confirmed by the inhibition of sneezing, pruritus and rhinorrhea by H_1R antagonists following nasal challenge with histamine.[9-13] H_2R antagonists have no effect on histamine-induced neuronal symptoms.[9,12,14]

Both pruritus and sneezing are completely mediated by nasal sensory nerves that originate from the 5th cranial nerve (trigeminal nerve). These afferent sensory nerves (which are distinct from olfactory nerves, 1st cranial nerve, that encode information of smell) innervate the epithelium, blood vessels and glands of the nasal mucosa. Although detailed studies in this field are limited, it appears that there are at least two kinds of nerves innervating the nasal mucosa of mammals:[15] touch/air flow-sensitive fibers and neuropeptide-containing polymodal nociceptors which respond to noxious stimuli such as histamine, ATP, mustard oil and capsaicin.[16-18] It is likely that this nociceptive subtype mediates allergen and histamine-induced sneezing and pruritus.[19-22] Nasal trigeminal afferents terminate in the brain stem trigeminal nuclei, forming synapses with central neurons that either ascend to the thalamus (sensations) or modulate nasal and lower airway efferent nerve activity (central reflexes). Histamine is unable to directly activate nasal submucosal glands,[23] but histamine-induced activation of nasal nociceptive sensory nerves causes a central reflex-dependent bilateral increase in the release of acetylcholine from nasal parasympathetic nerves, which results in increased mucus secretion from nasal submucosal glands.[24] Correspondingly, contralateral glandular secretion induced by unilateral histamine challenge is only reduced by muscarinic antagonists that are applied to the contralateral side, unlike H_1R antagonists that are effective when applied to the ipsilateral side.[25] Glandular secretion and plasma extravasation (see below) combine to cause the sensation of rhinorrhea associated with histamine challenge and may contribute to the sensation of nasal congestion.

Allergen-induced sneezing, pruritus and rhinorrhea in allergic individuals are reduced by H_1R antagonists (Table 1), but not by H_2R antagonists.[26,27] Evidence suggests that allergen-induced sneezing is largely abolished whereas residual itch and rhinorrhea remains in the presence of H_1R antagonists. The reasons for this remain unknown, but it is known that some nociceptor stimuli, such as capsaicin (pungent ingredient in chilli peppers) can cause reflex glandular activation without eliciting sneeze.[28] Perhaps histamine is the sole allergen-induced mediator that contributes to the initiation of sneeze, but other mediators are released during allergic inflammation that can activate nasal nociceptive nerves.

Table 1. Nasal blockage as measured subjectively by symptom scores (SS) and objectively by expiratory or inspiratory peak flow (PF), rhinomanometry (R) and acoustic rhinometry (AR)

Antagonist	Allergen Exposure	Sneeze	Pruritus	Rhinorrhea	Nasal Blockage	Ref
Astemizole	Challenge	↓	↓		↓(R)	66
Astemizole	Ambient	↓		↓	No effect (SS)	67
Chlorpheniramine	Ambient	↓	↓	↓	No effect (SS)	68
Cetirizine	Ambient	↓	↓	↓		69
Cetirizine	Ambient	↓	↓	↓	No effect (SS)	70
Cetirizine	Challenge	↓	↓	↓	No effect (AR)	11
Cetirizine	Challenge	↓	↓	↓	No effect (PF)	71
Desloratadine	Ambient		↓	No effect	↓(PF)	72
Desloratadine	Challenge	↓		↓	↓(PF,SS)	73
Desloratadine	Ambient				↓(SS)	74
Desloratadine	Challenge				No effect (PF)	60
Desloratadine	Ambient	↓	No effect	No effect	No effect (SS)	75
Desloratadine	Ambient	↓	↓	↓	No effect (SS), ↓(R)	76
Desloratadine	Ambient	↓	↓	↓	No effect (SS)	77
Diphenhydramine	Ambient	↓	↓	↓	↓(SS)	75
Dimetindene	Challenge	↓	↓	↓	No effect (PF)	78
Fexofenadine	Ambient		↓	No effect	↓(PF)	72
Hydroxyzine	Challenge	↓	↓	↓	No effect (SS)	79
Hydroxyzine	Ambient	↓	↓	↓	No effect (SS)	68
Levocabastine	Challenge				No effect (PF)	80
Levocabastine	Challenge	↓		↓	No effect (R)	81
Levocabastine	Challenge	↓		↓	No effect (R,SS)	26
Olopatadine	Challenge	↓	↓	↓	↓(SS)	82
Terfenadine	Challenge	↓		↓	No effect (R)	83
Terfenadine	Ambient	↓		No effect	No effect (SS)	67

In addition to its effects on nasal reflexes, there is evidence that histamine-induced activation of nasal sensory nerves can augment cough sensitivity in some individuals. Although histamine nasal challenge does not evoke cough, it increases the number of coughs elicited by 'mouth-only' inhalation of the tussive stimulant capsaicin.[29] Cough associated with allergic rhinitis remains a controversial topic,[2] but H_1R antagonists are commonly used as a first line of defense in the treatment of chronic cough in individuals with allergic rhinitis. The evidence suggests that 1st generation H_1R antagonists (e.g., chlorpheniramine) are perhaps better at reducing cough compared to 2nd generation H_1R antagonists (e.g., cetirizine), although the mechanisms through which this occurs is not fully understood.[3]

H_1 Receptors and the Nasal Vasculature

Nasal blockage is defined as a parameter objectively measured using acoustic rhinometry, rhinomanometry or nasal peak flow meters and is thought to be the result of vasodilation of nasal blood vessels, in particular sinusoids (also known as capacitance vessels). This causes the turbinate tissue to protrude into the nasal cavity, which is experienced as nasal congestion. Initially histamine was thought to cause nasal blockage solely through the activation of H_1R. The selective H_1R agonist betahistine causes nasal blockage and H_1R antagonists reduce histamine-induced nasal blockage.[8-11,30,31] H_1R have been shown, using autoradiography, on human turbinate vascular endothelium[32] and histamine induces an increase in calcium in cultured human turbinate endothelial cells that is blocked by H_1R antagonists but not H_2R or H_3R antagonists.[33] In addition, in vitro airway blood vessels fail to dilate in response to histamine following endothelial denudation.[34] Human studies with nitric oxide synthase[35] and cyclooxygenase inhibitors[36] suggest that neither nitric oxide nor prostanoids are the H_1R-induced endothelium-dependent relaxing factor. However, nitric oxide does play a role in H_1R-mediated increase in plasma extravasation.[35] Through this mechanism barrier function in the nasal vasculature becomes leaky, resulting in increased plasma release into the nasal cavity (which contributes to rhinorrhea).

H_1R antagonists, however, have been far less effective in reducing allergen-induced nasal blockage (Table 1). A large number of studies have been published investigating the effect of a range of H_1R antagonists in reducing nasal blockage as measured by a wide variety of objective and subjective techniques. Many studies failed to demonstrate a reduction in allergen-induced nasal blockage. Meta-analysis by Hore et al (2005) of double-blind randomized controlled trials of oral H_1R antagonists suggest that there is indeed a significant reduction of allergen-induced nasal blockage,[37] but this seems at best to be inconsistently demonstrated throughout the literature (at least compared to the effectiveness of H_1R antagonists in reducing allergen-induced sneezing). One H_1R antagonist that may be efficacious is desloratadine. However, compared to other H_1R antagonists such as cetirizine and fexofenadine, desloratadine is relatively unselective for H_1R.[38] Calculations based on their published H_1R affinities and oral bioavailability suggest that both desloratadine and cetirizine occupy a very similar proportion of H_1R following standard oral doses. Thus it seems likely if desloratadine does reduce allergen-induced nasal blockage, it does so via nonH_1R-mediated mechanisms. The failure of H_1R antagonists to reduce allergen-induced nasal blockage is unlikely to be due of a decrease in H_1R on nasal blood vessels, as H_1R mRNA levels are higher in individuals with allergic rhinitis.[39] More likely is the existence of redundancy in the vasodilatory mechanisms following allergen challenge.

Other Histamine Receptors in the Early Phase Response

Analysis of the effect of high-dose H_1R antagonists on histamine-induced nasal blockage suggests that histamine causes nasal blockage via H_1R and nonH_1R mediated pathways.[31] Consistent with this, dimaprit, the selective H_2R agonist, causes nasal blockage in humans that is abolished by H_2R antagonists.[8,31] Based on work in other vascular systems, the H_2R-mediated nasal blockage is likely due to the activation of H_2R on the nasal vascular smooth muscle, causing endothelial-independent vasodilation.[40] Despite this data, histamine-induced nasal blockage is inconsistently reduced by H_2R antagonists, although combinations of H_1R and H_2R antagonists appear to be more effective than H_1R antagonists alone.[6] Importantly, in the limited number of published studies, the combination of H_1R and H_2R antagonists failed to reduce allergen-induced nasal blockage.[26,41]

Nasal challenge with the selective H_3R agonist R-α-methylhistamine causes nasal blockage in humans that is reduced by a topical application of the H_3R/H_4R antagonist thioperamide.[31] In vitro mechanistic studies of nasal tissue taken from both pigs and humans demonstrate that R-α-methylhistamine indirectly causes vasodilation by reducing the release of the vasoconstrictor noradrenaline from nasal sympathetic nerve endings.[42,43] Compound 48/80, a mast cell activator, mimicked the effect of R-α-methylhistamine and both were inhibited by thioperamide (H_3R/H_4R) and clobenpropit (H_3R antagonist/H_4R agonist). Nasal blood vessels and, to a much lesser extent, submucosal glands are innervated by sympathetic neurons originating from the superior cervical ganglion. Stimulation of nasal sympathetic nerves in vitro causes vasoconstriction mainly via noradrenaline's activation of α_1-adrenoceptors on the vascular smooth muscle,[42,44] although postjunctional α_2-adrenoceptors and neuropeptide Y may also play a role.[45,46] Consistent with these in vitro studies, nasal challenge with the selective α_1-adrenoceptor agonist corynanthine caused nasal blockage in humans,[31] demonstrating that there is a basal sympathetic vasoconstricting tone in vivo.

In vivo evidence for a contribution of H_3R in histamine-induced nasal blockage has been demonstrated in both cats and humans. However, the H_3R antagonist thioperamide only reduced histamine-induced nasal blockage in humans when given in combination with an H_1R antagonist.[31] Similar results have been observed in compound 48/80-induced nasal blockage in cats.[47] The reasons for this requirement of H_1R antagonism are not clear, but one theory suggests that activation of H_1R receptors functionally antagonizes the effect of H_3R activation on sympathetic output.[6]

Unfortunately, the H_3R antagonists available at the moment (thioperamide and clobenpropit) are not suitable for clinical development and as such we must await more potent and selective tools before any assessment of the role of H_3R in allergen-induced nasal blockage in humans. If allergen-induced histamine release does indeed reduce the sympathetic tone to the nasal blood vessels via H_3R, perhaps a combination of H_1R and H_3R antagonists may be efficacious. Indeed the function of efferent autonomic nerves in the nasal mucosa may be perverted in allergic individuals,[45,48] which may result in a more critical role for H_3R. However, it is important to note that in murine models of allergic rhinitis a different role of H_3R has been described.[49-51] In these studies a combination of an H_1R antagonist and an H_3R agonist reduced allergic symptoms. The mechanism suggested for the role of H_3R was through the inhibition of substance P release from nasal nociceptive sensory nerves. Although H_3R have been immunohistochemically demonstrated on human nasal nerves,[7] this study unfortunately did not characterize these nerve fibers (e.g., afferent vs efferent autonomic). Nevertheless, it seems unlikely that a reduction in substance P release from nociceptive sensory nerves in the human nasal mucosa would greatly inhibit allergen-induced sneeze, pruritus, rhinorrhea and nasal congestion because although human nasal challenge with substance P does cause plasma extravasation (and therefore rhinorrhea) and nasal blockage, it does not cause sneeze or pruritus.[52,53] In addition, nasal challenge with capsaicin, which causes the activation of nociceptive sensory nerves innervating the airways,[16] leading to the release of substance P from afferent peripheral terminals, fails to induce nasal blockage,[28,54,55] suggesting that endogenously produced substance P is unable to elicit the nasal responses that exogenously applied substance P can. Unfortunately, until better pharmacological tools (specifically selective and potent H_3R ligands) can be developed, the role of H_3R in the early phase response of allergic rhinitis will remain unclear.

Histamine Receptors and Immune Modulation

Many immune cell types including mast cells, eosinophils, dendritic cells, T-lymphocytes, monocytes and macrophages, express histamine receptors and there is growing evidence that histamine can exert influence on the immune response. In some cases multiple histamine receptors are expressed on the same cell type and activation of these receptors can have opposing effects to one another. Histamine modulates chemotaxis of granulocytes: positively via H_1R and negatively via H_2R.[56] Histamine also modulates the production of Th2 cytokines and chemokines, mainly via the H_1R.[57] There is some evidence that H_1R antagonists reduce mast cell function and eosinophilia following allergen challenge in allergic subjects,[58-62] but there is ongoing debate on whether these effects were due to off-target functions of the antihistamine.[63-65]

Conclusion

Histamine is released during both early and late phases of the nasal allergic response and can modulate the function of virtually every important cell type in the nasal mucosa. The principle pro-rhinitis actions of histamine are mediated via H_1R. H_2R appear to play a minor role. The discovery of H_3R and H_4R have yielded interesting and sometimes controversial hypotheses, but until selective and safe ligands can be developed and tested in the clinic, the debate will rage on.

References

1. Skoner DP. Allergic rhinitis: Definition, epidemiology, pathophysiology, detection and diagnosis. Journal of Allergy and Clinical Immunology 2001; 108:S2-S8.
2. Morice AH. Epidemiology of cough. In: Chung F, Widdicombe J, Boushey HA, eds. Cough: Causes, Mechanisms and Therapy: Blackwell 2003:11-6.
3. Bolser DC. Older-generation antihistamines and cough due to upper airway cough syndrome (UACS): efficacy and mechanism. Lung 2008; 186(Suppl 1):S74-7.
4. Doyle WJ, Boehm S, Skoner DP. Physiologic responses to intranasal dose-response challenges with histamine, methacholine, bradykinin and prostaglandin in adult volunteers with and without nasal allergy. J Allergy Clin Immunol 1990; 86:924-35.
5. Austin CE, Foreman JC. Acoustic rhinometry compared with posterior rhinomanometry in the measurement of histamine- and bradykinin-induced changes in nasal airway patency. Br J Clin Pharmacol 1994; 37:33-7.
6. Taylor-Clark T, Foreman J. Histamine-mediated mechanisms in the human nasal airway. Curr Opin Pharmacol 2005; 5:214-20.
7. Nakaya M, Takeuchi N, Kondo K. Immunohistochemical localization of histamine receptor subtypes in human inferior turbinates. Ann Otol Rhinol Laryngol 2004; 113:552-7.
8. Shelton D, Eiser N. Histamine receptors in the human nose. Clin Otolaryngol Allied Sci 1994; 19:45-9.
9. Secher C, Kirkegaard J, Borum P et al. Significance of H1 and H2 receptors in the human nose: rationale for topical use of combined antihistamine preparations. J Allergy Clin Immunol 1982; 70:211-8.
10. Kirkegaard J, Secher C, Borum P et al. Inhibition of histamine-induced nasal symptoms by the H1 antihistamine chlorpheniramine maleate: demonstration of topical effect. Br J Dis Chest 1983; 77:113-22.
11. Hilberg O, Grymer LF, Pedersen OF. Nasal histamine challenge in nonallergic and allergic subjects evaluated by acoustic rhinometry. Allergy 1995; 50:166-73.
12. Wood-Baker R, Lau L, Howarth PH. Histamine and the nasal vasculature: the influence of H1 and H2-histamine receptor antagonism. Clin Otolaryngol Allied Sci 1996; 21:348-52.
13. Wang DY, Hanotte F, De Vos C et al. Effect of cetirizine, levocetirizine and dextrocetirizine on histamine-induced nasal response in healthy adult volunteers. Allergy 2001; 56:339-43.
14. Mygind N, Secher C, Kirkegaard J. Role of histamine and antihistamines in the nose. Eur J Respir Dis Suppl 1983; 128(Pt 1):16-20.
15. Canning BJ. Neurology of allergic inflammation and rhinitis. Curr Allergy Asthma Rep 2002; 2:210-5.
16. Taylor-Clark TE, Kollarik M, MacGlashan DW, Jr, et al. Nasal sensory nerve populations responding to histamine and capsaicin. J Allergy Clin Immunol 2005; 116:1282-8.
17. Damann N, Rothermel M, Klupp BG et al. Chemosensory properties of murine nasal and cutaneous trigeminal neurons identified by viral tracing. BMC Neurosci 2006; 7:46.
18. Taylor-Clark TE, Kiros F, Carr MJ et al. Transient Receptor Potential Ankyrin 1 Mediates Toluene Diisocyanate-Evoked Respiratory Irritation. Am J Respir Cell Mol Biol 2008.
19. Asakura K, Narita S, Kojima T et al. Changes in nasal airway resistance and secretory response in the guinea pig after nasal challenge with capsaicin and histamine. Eur Arch Otorhinolaryngol 1994; 251:224-8.
20. Asakura K, Narita S, Kojima T et al. Role of capsaicin-sensitive sensory nerve reflexes in guinea pig model of nasal allergy. Int Arch Allergy Immunol 1993; 102:195-9.
21. Blom HM, Van Rijswijk JB, Garrelds IM et al. Intranasal capsaicin is efficacious in non-allergic, non-infectious perennial rhinitis. A placebo-controlled study. Clin Exp Allergy 1997; 27:796-801.
22. Van Rijswijk JB, Boeke EL, Keizer JM et al. Intranasal capsaicin reduces nasal hyperreactivity in idiopathic rhinitis: a double-blind randomized application regimen study. Allergy 2003; 58:754-61.
23. Mullol J, Raphael GD, Lundgren JD et al. Comparison of human nasal mucosal secretion in vivo and in vitro. J Allergy Clin Immunol 1992; 89:584-92.

24. Naclerio RM, Baroody FM. Response of nasal mucosa to histamine or methacholine challenge: use of a quantitative method to examine the modulatory effects of atropine and ipratropium bromide. J Allergy Clin Immunol 1992; 90:1051-4.
25. Baroody FM, Wagenmann M, Naclerio RM. Comparison of the secretory response of the nasal mucosa to methacholine and histamine. J Appl Physiol 1993; 74:2661-71.
26. Holmberg K, Pipkorn U, Bake B et al. Effects of topical treatment with H1 and H2 antagonists on clinical symptoms and nasal vascular reactions in patients with allergic rhinitis. Allergy 1989; 44:281-7.
27. Wang D, Clement P, Smitz J. Effect of H1 and H2 antagonists on nasal symptoms and mediator release in atopic patients after nasal allergen challenge during the pollen season. Acta Otolaryngol 1996; 116:91-6.
28. Philip G, Baroody FM, Proud D et al. The human nasal response to capsaicin. J Allergy Clin Immunol 1994; 94:1035-45.
29. Plevkova J, Brozmanova M, Pecova R et al. The effects of nasal histamine challenge on cough reflex in healthy volunteers. Pulm Pharmacol Ther 2006; 19:120-7.
30. Havas TE, Cole P, Parker L et al. The effects of combined H1 and H2 histamine antagonists on alterations in nasal airflow resistance induced by topical histamine provocation. J Allergy Clin Immunol 1986; 78:856-60.
31. Taylor-Clark T, Sodha R, Warner B et al. Histamine receptors that influence blockage of the normal human nasal airway. Br J Pharmacol 2005.
32. Ishibe T, Kubo N, Kumazawa H et al. Histamine H1 receptors and affinity analyses in human nasal mucosa in cases of nasal allergy. Ann Otol Rhinol Laryngol 1985; 94:186-90.
33. Ikeda H, Kubo N, Nakamura A et al. Histamine-induced calcium released from cultured human mucosal microvascular endothelial cells from nasal inferior turbinate. Acta Otolaryngol 1997; 117:864-70.
34. Ortiz JL, Labat C, Norel X et al. Histamine receptors on human isolated pulmonary arterial muscle preparations: effects of endothelial cell removal and nitric oxide inhibitors. J Pharmacol Exp Ther 1992; 260:762-7.
35. Dear JW, Ghali S, Foreman JC. Attenuation of human nasal airway responses to bradykinin and histamine by inhibitors of nitric oxide synthase. Br J Pharmacol 1996; 118:1177-82.
36. McLean JA, Bacon JR, Mathews KP et al. Effects of aspirin on nasal responses in atopic subjects. J Allergy Clin Immunol 1983; 72:187-92.
37. Hore I, Georgalas C, Scadding G. Oral antihistamines for the symptom of nasal obstruction in persistent allergic rhinitis—a systematic review of randomized controlled trials. Clin Exp Allergy 2005; 35:207-12.
38. Gillard M, Christophe B, Wels B et al. H1 antagonists: receptor affinity versus selectivity. Inflamm Res 2003; 52(Suppl 1):S49-50.
39. Iriyoshi N, Takeuchi K, Yuta A et al. Increased expression of histamine H1 receptor mRNA in allergic rhinitis. Clin Exp Allergy 1996; 26:379-85.
40. Chang JY, Hardebo JE, Owman C. Differential vasomotor action of noradrenaline, serotonin and histamine in isolated basilar artery from rat and guinea-pig. Acta Physiol Scand 1988; 132:91-102.
41. Juliusson S, Bende M. Effect of systemically administered H1- and H2-receptor antagonists on nasal blood flow as measured with laser Doppler flowmetry in a provoked allergic reaction. Rhinology 1996; 34:24-7.
42. Varty LM, Gustafson E, Laverty M et al. Activation of histamine H3 receptors in human nasal mucosa inhibits sympathetic vasoconstriction. Eur J Pharmacol 2004; 484:83-9.
43. Varty LM, Hey JA. Histamine H3 receptor activation inhibits neurogenic sympathetic vasoconstriction in porcine nasal mucosa. Eur J Pharmacol 2002; 452:339-45.
44. Johannssen V, Maune S, Werner JA et al. Alpha 1-receptors at precapillary resistance vessels of the human nasal mucosa. Rhinology 1997; 35:161-5.
45. Malm L, Okuda M, Dieges PH. Direct and reflex actions of biochemical mediators. In: Allergic and Vasomotor Rhinitis: Pathophysiological Aspects 1987:214-9.
46. Lacroix JS, Stjarne P, Anggard A et al. Sympathetic vascular control of the pig nasal mucosa (2): Reserpine-resistant, non-adrenergic nervous responses in relation to neuropeptide Y and ATP. Acta Physiol Scand 1988; 133:183-97.
47. McLeod RL, Mingo GG, Herczku C et al. Combined histamine H1 and H3 receptor blockade produces nasal decongestion in an experimental model of nasal congestion. Am J Rhinol 1999; 13:391-9.
48. Sanico AM, Philip G, Proud D et al. Comparison of nasal mucosal responsiveness to neuronal stimulation in non-allergic and allergic rhinitis: effects of capsaicin nasal challenge. Clin Exp Allergy 1998; 28:92-100.

49. Nakaya M, Fukushima Y, Takeuchi N et al. Nasal allergic response mediated by histamine H3 receptors in murine allergic rhinitis. Laryngoscope 2005; 115:1778-84.
50. Yokota E, Kuyama S, Ogawa M et al. Substance P is involved in the effect of histamine H3 receptor agonist, Sch 50971 on nasal allergic symptoms in mice. Int Immunopharmacol 2008; 8:1083-8.
51. Yokota E, Kuyama S, Sugimoto Y et al. Participation of histamine H3 receptors in experimental allergic rhinitis of mice. J Pharmacol Sci 2008; 108:206-11.
52. Devillier P, Dessanges JF, Rakotosihanaka F et al. Nasal response to substance P and methacholine in subjects with and without allergic rhinitis. Eur Respir J 1988; 1:356-61.
53. Braunstein G, Fajac I, Lacronique J et al. Clinical and inflammatory responses to exogenous tachykinins in allergic rhinitis. Am Rev Respir Dis 1991; 144:630-5.
54. Rajakulasingam K, Polosa R, Lau LC et al. Nasal effects of bradykinin and capsaicin: influence on plasma protein leakage and role of sensory neurons. J Appl Physiol 1992; 72:1418-24.
55. Sanico AM, Atsuta S, Proud D et al. Dose-dependent effects of capsaicin nasal challenge: in vivo evidence of human airway neurogenic inflammation. J Allergy Clin Immunol 1997; 100:632-41.
56. Akdis CA, Simons FE. Histamine receptors are hot in immunopharmacology. Eur J Pharmacol 2006; 533:69-76.
57. Schneider E, Rolli-Derkinderen M, Arock M et al. Trends in histamine research: new functions during immune responses and hematopoiesis. Trends Immunol 2002; 23:255-63.
58. Fadel R, David B, Rassemont R et al. Eosinophil infiltration: effects of H1 antihistamines. J Am Acad Dermatol 1991; 24:1094-6.
59. Ciprandi G, Cirillo I, Vizzaccaro A et al. Desloratadine and levocetirizine improve nasal symptoms, airflow and allergic inflammation in patients with perennial allergic rhinitis: a pilot study. Int Immunopharmacol 2005; 5:1800-8.
60. Reinartz SM, Overbeek SE, Kleinjan A et al. Desloratadine reduces systemic allergic inflammation following nasal provocation in allergic rhinitis and asthma patients. Allergy 2005; 60:1301-7.
61. Mahmoud F, Arifhodzic N, Haines D et al. Levocetirizine modulates lymphocyte activation in patients with allergic rhinitis. J Pharmacol Sci 2008; 108:149-56.
62. Lee CF, Sun HL, Lu KH et al. The comparison of cetirizine, levocetirizine and placebo for the treatment of childhood perennial allergic rhinitis. Pediatr Allergy Immunol 2008.
63. Mygind N. Nasal inflammation and anti-inflammatory treatment. Semantics or clinical reality. Rhinology 2001; 39:61-5.
64. Assanasen P, Naclerio RM. Antiallergic anti-inflammatory effects of H1-antihistamines in humans. Clin Allergy Immunol 2002; 17:101-39.
65. Thurmond RL, Gelfand EW, Dunford PJ. The role of histamine H1 and H4 receptors in allergic inflammation: the search for new antihistamines. Nat Rev Drug Discov 2008; 7:41-53.
66. Horak F, Toth J, Jager S et al. Effects of H1-receptor antagonists on nasal obstruction in atopic patients. Allergy 1993; 48:226-9.
67. Howarth PH, Holgate ST. Comparative trial of two nonsedative H1 antihistamines, terfenadine and astemizole, for hay fever. Thorax 1984; 39:668-72.
68. Wong L, Hendeles L, Weinberger M. Pharmacologic prophylaxis of allergic rhinitis: relative efficacy of hydroxyzine and chlorpheniramine. J Allergy Clin Immunol 1981; 67:223-8.
69. Falliers CJ, Brandon ML, Buchman E et al. Double-blind comparison of cetirizine and placebo in the treatment of seasonal rhinitis. Ann Allergy 1991; 66:257-62.
70. DuBuske L. Dose-ranging comparative evaluation of cetirizine in patients with seasonal allergic rhinitis. Ann Allergy Asthma Immunol 1995; 74:345-54.
71. Korsgren M, Andersson M, Borga O et al. Clinical efficacy and pharmacokinetic profiles of intranasal and oral cetirizine in a repeated allergen challenge model of allergic rhinitis. Ann Allergy Asthma Immunol 2007; 98:316-21.
72. Wilson AM, Haggart K, Sims EJ et al. Effects of fexofenadine and desloratadine on subjective and objective measures of nasal congestion in seasonal allergic rhinitis. Clin Exp Allergy 2002; 32:1504-9.
73. Horak F, Stubner UP, Zieglmayer R et al. Effect of desloratadine versus placebo on nasal airflow and subjective measures of nasal obstruction in subjects with grass pollen-induced allergic rhinitis in an allergen-exposure unit. J Allergy Clin Immunol 2002; 109:956-61.
74. Nayak AS, Schenkel E. Desloratadine reduces nasal congestion in patients with intermittent allergic rhinitis. Allergy 2001; 56:1077-80.
75. Raphael GD, Angello JT, Wu MM et al. Efficacy of diphenhydramine vs desloratadine and placebo in patients with moderate-to-severe seasonal allergic rhinitis. Ann Allergy Asthma Immunol 2006; 96:606-14.
76. Meltzer EO, Jalowayski AA, Vogt K et al. Effect of desloratadine therapy on symptom scores and measures of nasal patency in seasonal allergic rhinitis: results of a single-center, placebo-controlled trial. Ann Allergy Asthma Immunol 2006; 96:363-8.

77. Pradalier A, Neukirch C, Dreyfus I et al. Desloratadine improves quality of life and symptom severity in patients with allergic rhinitis. Allergy 2007; 62:1331-4.
78. Horak F, Unkauf M, Beckers C et al. Efficacy and tolerability of intranasally applied dimetindene maleate solution versus placebo in the treatment of seasonal allergic rhinitis. Arzneimittelforschung 2000; 50:1099-105.
79. Brooks CD, Nelson A, Parzyck R et al. Protective effect of hydroxyzine and phenylpropanolamine in the challenged allergic nose. Ann Allergy 1981; 47:316-9.
80. Corren J, Rachelefsky G, Spector S et al. Onset and duration of action of levocabastine nasal spray in atopic patients under nasal challenge conditions. J Allergy Clin Immunol 1999; 103:574-80.
81. Pecoud A, Zuber P, Kolly M. Effect of a new selective H1 receptor antagonist (levocabastine) in a nasal and conjunctival provocation test. Int Arch Allergy Appl Immunol 1987; 82:541-3.
82. Pipkorn P, Costantini C, Reynolds C et al. The effects of the nasal antihistamines olopatadine and azelastine in nasal allergen provocation. Ann Allergy Asthma Immunol 2008; 101:82-9.
83. Rokenes HK, Andersson B, Rundcrantz H. Effect of terfenadine and placebo on symptoms after nasal allergen provocation. Clin Allergy 1988; 18:63-9.

Chapter 5

The Role of Histamine in Ocular Allergy

Masaharu Ohbayashi, Bita Manzouri, Kei Morohoshi, Ken Fukuda and Santa J. Ono*

Abstract

Ocular allergy is a disorder affecting increasing numbers of individuals worldwide. Among the inflammatory mediators that contribute to ocular allergy, histamine is perhaps the best characterized. This monoamine is released by sensitized mast cells upon exposure to allergen and causes symptoms such as redness and tearing. Histamine may also recruit immune cells that can cause long-term damage to ocular surfaces. In this chapter we will discuss the known functions of histamine and histamine receptors in ocular allergy and will describe promising therapies targeting the histamine-signaling pathway.

Introduction

Allergic diseases are widely prevalent with one study showing about 35% of families interviewed in the United States experiencing some form of allergy and more than 50% of those affected reporting eye symptoms.[1] Worldwide, allergies are estimated to affect 20% of the population.[2] Allergic pathogenesis is therefore the focus of intensive research and many studies have focused on the role of histamine in ocular allergy and other atopic disorders.

Histamine is a biogenic amine that plays a key role in allergic inflammation. This molecule is closely associated with immunoglobulin E (IgE)-mediated mast cell activation in conjunctival tissues.[3] It also regulates physiological functions in the gut and can act as a neurotransmitter.[4] As part of an immune response to foreign pathogens, histamine is produced by basophils and mast cells found in the local connective tissue. Upon release it acts to increase the permeability of the capillaries to other white blood cells.[5] These changes in vascular permeability along with other effects of histamine play critical roles in the pathogenesis of ocular allergy.

Clinical Manifestations of Ocular Allergy

The manifestations of allergic eye disease comprise a heterogeneous group of clinical conditions that range from simple, intermittent symptoms of itching, tearing and redness to severe, sight-threatening corneal complications. These conditions may be considered as part of an immunologic spectrum—incorporating both Type I and Type IV hypersensitivities—that affect the anterior surface of the eye.[6] Allergic eye diseases have a degree of overlap, but are traditionally classified into five distinct entities (Table 1): seasonal allergic conjunctivitis (SAC), perennial allergic conjunctivitis (PAC), vernal keratoconjunctivitis (VKC), atopic keratoconjunctivitis (AKC) and giant papillary conjunctivitis (GPC). GPC is an iatrogenic disease that is associated with foreign bodies on the eye (e.g., contact lenses, ocular prostheses) and is not always included in this grouping. GPC invariably resolves when the cause is removed and keratopathy is rare. It was thought

*Corresponding Author: Santa J. Ono—Office of the Provost, Emory University. Administration Building, Suite 404, Atlanta, Georgia 30322, USA. Email: sjono@emory.edu

Histamine in Inflammation, edited by Robin L. Thurmond.

Table 1. The spectrum of allergic eye disease

	Hypersensitive Type	Sight-Threatening?
Seasonal allergic conjunctivitis	Type I response to seasonal allergen	no
Perennial allergic conjunctivitis	Type I response to seasonal allergen	no
Vernal keratoconjunctivitis	Type I and Type IV hypersensitivity (Chronic), sometimes associated with other atopic conditions	yes
Atopic keratoconjunctivitis	Chronic, often associated with atopic dermatitis	yes
Giant papillary conjunctivitis	Mast cell-mediated or inflammatory	rarely

to have a possible allergic mechanism because of the predominance of mast cells, but now the underlying mechanism of GPC is thought to be an inflammatory rather than an allergic response.[1]

Seasonal and Perennial Allergic Conjunctivitis

SAC (also known as hay fever conjunctivitis) and PAC are the most common ocular allergic disorders worldwide. Although neither results in blindness, they can cause severe incapacitation in affected patients. Both are bilateral, self-limiting conjunctival inflammatory processes that occur in susceptible individuals of either sex. The condition is initiated by either a seasonal (e.g., ragweed, birch or grass pollen) or a perennial (e.g., cat dander, house dust mite) allergen binding to IgE on mast cells and triggering the Type I hypersensitivity response. It is this response that leads to the clinical symptoms of itching and watering eyes that are associated with conjunctival redness and edema. Neither SAC nor PAC causes permanent ocular surface damage.

Vernal Keratoconjunctivitis

VKC is a sight-threatening, bilateral, chronic inflammatory process that mainly affects males. The onset tends to be early in life, usually before the age of 5 and resolution generally occurs by the end of puberty.[1] Individuals who live in warm, dry climates tend to be more commonly afflicted, with the Mediterranean and Western Africa having the greatest numbers of patients. About three-quarters of VKC patients tend to have a significant history of atopy, often also suffering from asthma or eczema and a positive family history of atopy is found in two-thirds of patients.[1] Symptoms are similar to those of SAC and PAC and include pain, itching, conjunctival injection and mucous discharge. Although the symptoms may be present year-round, patients tend to have seasonal exacerbations. Clinical signs include superior tarsal conjunctival papillae, conjunctival hyperemia and edema. Horner-Trantas dots, which are composed of clumps of dead epithelial cells and eosinophils, may be found on the superior limbus. Corneal involvement in VKC with the development of ulceration can lead to sight-threatening complications, such as corneal scarring.

Atopic Keratoconjunctivitis

AKC is another bilateral chronic inflammatory disease of the conjunctiva that has the potential to be sight threatening and, unlike VKC, it has a very strong association with atopic dermatitis (eczema). Atopic dermatitis, a pruritic skin condition that affects 3% of the population worldwide, is present in 95% of patients with AKC.[7] Conversely, 25-40% of atopic dermatitis patients have AKC.[8] The onset of the disease is usually between the ages of 20 and 50 years and there seems to be a slight male preponderance. Symptoms are similar to VKC, but patients may also complain of a mucous discharge, especially upon wakening in the morning. Clinical signs are also similar to VKC, with the exception of papillary hypertrophy, which may be present on the lower tarsal conjunctiva in AKC. In addition, these patients have variable signs of disease on the lid margins (thickening of the margin, meiobomian gland dysfunction, blepharitis with crusts) and the periorbital skin (dry and scaly skin, an extra skin fold of the lower eye lid and fissures of the lateral canthus of the eye).

Current Treatments for Ocular Allergy

The basic aim in treating allergic eye disease is to achieve relief from symptoms and control of the underlying condition. Additionally, in the more severe forms of allergy such as VKC and AKC, the aim also extends to preventing visual complications that may result, the most severe of which is sight loss. Basic treatment—applicable to all forms of allergic eye disease—is aimed at removing or diluting the allergens that may come into contact with the ocular surface. Patients are advised allergen avoidance if the inciting allergen is known, and the use of cold compresses is recommended for symptomatic relief of the pruritus experienced in these conditions. It is further recommended that patients keep their ocular medications refrigerated so that additional symptomatic relief can be achieved when the cold drops are applied to the ocular surface. Lubrication, in the form of tear substitutes, are used to dilute the allergen and/or wash out the inflammatory mediators from the ocular surface, thereby providing symptomatic relief. Lubricating ointments can be applied to the patient's eye just before they go to sleep and can provide moisture to the ocular surface overnight.

Given the prevalence of mast cells in the conjunctiva in all forms of allergic conjunctivitis (see below), it is not surprising that mast-cell stabilizers and antihistamines form the mainstay of topical treatments. Dual-acting agents, combining antihistamine effects with inhibition of mediator release, are the newest generation of anti-allergic agents. They offer the advantage of rapid relief of symptoms coupled with the long-term disease modifying benefits. In the more severe or chronic forms, additional therapy with topical anti-inflammatory treatment is also often required, usually in the form of topical corticosteroids. Topical steroid preparations are the most effective therapy for use in moderate to severe forms of allergic eye disease. However, their use should be strictly limited to severe cases and the patient will require careful monitoring by an ophthalmologist. Long-term use of these drugs is associated with an increased risk of cataract formation and glaucoma and they can potentiate ocular herpetic infections.

The calcineurin inhibitors cyclosporine A and tacrolimus, drugs used to induce systemic immunosuppression following organ transplantation, can also be effective in treating ocular allergy. The enzyme calcineurin plays an important role in cell receptor signaling and is inactivated by cyclosporine A. In addition, cyclosporine A inhibits histamine release from mast cells.[9] Systemic cyclosporine A has been shown to improve symptoms in severe AKC,[10] but its use is associated with potentially life-threatening side effects such as renal failure, which makes it difficult to justify its use in nonlife-threatening diseases. Topical preparations of cyclosporine A circumvent the problem of systemic side effects and a 0.05% cyclosporine A emulsion formulation has been licensed in the United States for the treatment of dry eye disease. In a Phase III multicentre study this formulation proved to be safe and well tolerated.[11]

Tacrolimus has been approved for topical use in atopic dermatitis. It is available as an ointment in two strengths, 0.1% and 0.03% and has recently been reported to be effective in treating severe AKC and VKC.[12,13] There are a number of concerns regarding topical use of this drug, however. First, local immune deviation or suppression may increase a patient's susceptibility to local infections. Second, tacrolimus, even used topically, may be carcinogenic. The US FDA issued a public health advisory in March 2005 about the potential cancer risk and advised that tacrolimus should be used only as labeled, for patients in which other treatments have failed to work.[14]

Allergic Responses in the Conjunctiva

It is known that two cell types participate in the initiation of the ocular allergic response. Mast cells, which are perhaps the key cellular component of ocular allergy, are found in increased numbers in all forms of allergic conjunctivitis. The second cell type is the conjunctival T cell, whose numbers, particularly of the $CD4^+$ memory cell subtype, are simultaneously increased with those of mast cells in VKC and AKC.[15,16] The pathophysiology of allergic eye disease involves a Type I and/or IV hypersensitivity reaction. The acute forms of allergic eye disease (i.e., SAC and PAC) typically involve a mast cell-mediated Type I hypersensitivity reaction, whereas in the more chronic conditions of AKC and VKC, a Type IV hypersensitivity response is also thought to be involved.

Ocular allergic reactions typically occur in three phases: the sensitization phase, the early phase and the late phase. The sensitization phase occurs upon initial exposure of the ocular surface to aeroallergens. These allergens are phagocytosed by antigen-presenting cells such as dendritic cells on the mucosal epithelium of the conjunctiva. The allergens are processed within the antigen-presenting cells and then presented on the cell surface as a peptide fragment in association with the major histocompatibility complex class II molecule. This allergen-major histocompatibility complex on the surface of antigen-presenting cells then interacts with naïve $CD4^+$, or T helper (Th) cells, causing maturation of these naïve cells into Th type 1 (Th1) or Th type 2 (Th2) cells. It is Th2 cells that are mainly involved in the allergic response. The antigen-presenting cell-Th2 interaction results in the production of cytokines, which interact with naïve B cells and cause production of immunoglobulin E (IgE)-type antibodies. The IgE binds to its high-affinity receptor FcεRI on the surface of mast cells and basophils. When the sensitized eye subsequently encounters the same allergen, the allergen attaches to and cross-links IgE-FcεRI complexes on the surface of mast cells.[15] This leads to changes in the mast cell outer membrane, making the mast cell more permeable to calcium ions with subsequent mobilization of intracellular calcium. When a critical mass of IgE antibodies becomes cross-linked, this increased permeability causes the mast cells to degranulate, releasing a variety of primary inflammatory mediators that are stored in preformed granules. The mediators include histamine, serotonin, leukotriene C4, prostaglandin D2, platelet-activating factor, tryptase, chymase, cathepsin G and other eosinophil and neutrophil chemoattractants. These molecules are responsible for the symptoms of the "early-phase response", usually beginning within seconds of allergen contact and lasting for up to 40 min after exposure (Fig. 1). In ocular allergy, tear levels of histamine are increased during the early phase; ocular redness, tearing and itching also occur.[17]

A "late-phase reaction" sustained by a complex network of inflammatory cells and mediators can also occur in the eye. This has been demonstrated in humans using allergen for conjunctival provocation of allergic subjects.[18] Allergen challenge caused the typical early-phase reaction within 20 minutes, with the initial reaction being dose dependent. With smaller doses of allergen the reaction was not so pronounced and spontaneous recovery occurred within a brief period. With larger doses, the reaction was more persistent and progressed to a late-phase reaction. Typically, high doses of allergen induced a continuous response manifested by burning, redness, itching, tearing and a foreign body sensation that began 4-8 hours after challenge and persisted for up to 24 hours. This clinical reaction was accompanied by a significant recruitment of inflammatory cells in the tears: neutrophils first appeared about twenty minutes after challenge with eosinophils and lymphocytes increasing in prominence 6-24 hours after challenge.[18]

The eosinophil is the cell that predominates in the late phase reaction. It is a powerful effector cell, releasing arginine-rich toxic proteins that can damage the corneal epithelium.[15] Eosinophils are not normally found in the conjunctival epithelium of non-atopic subjects, but their numbers are increased in the conjunctival epithelium, subepithelium and tears of patients with AKC and VKC.

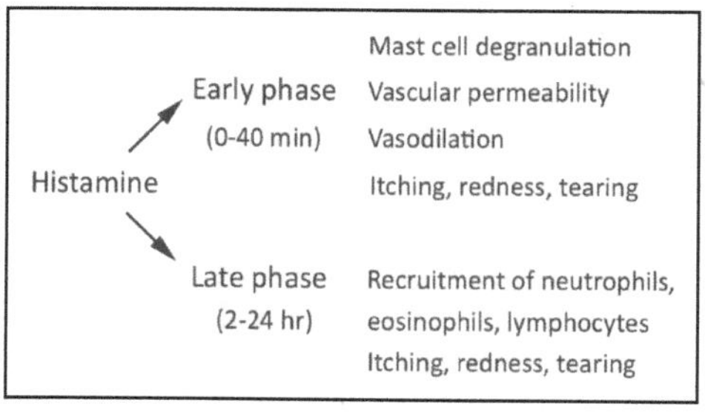

Figure 1. Histamine contributions to allergic responses.

In VKC, this increase in eosinophils and eosinophil products is observed in both skin test-positive and skin test-negative patients and is not confined to ocular tissues, suggesting that systemic activation of eosinophils may be more important than IgE activation in disease pathogenesis.[18]

Mast Cells in the Conjunctiva

Mast cells are subdivided into subtypes based on various physical and functional characteristics. The majority of conjunctival mast cells are of the MC_{TC} phenotype (with granules containing tryptase, chymase, carboxypeptidase and cathepsin) and are primarily found in the subepithelial region; the remaining mast cells are of the MC_T (with granules containing tryptase only) subtype.[19] Interestingly, mast cell subtype levels can vary depending on the type of ocular allergy. MC_T numbers increase in SAC, VKC and AKC, and in AKC the MC_{TC} subtype predominates.[19]

The healthy human conjunctiva has been estimated to contain approximately 11,000 mast cells/mm^3, found below the basement membrane in the substantia propria.[20] The number of mast cells in the conjunctiva increases in the chronic forms of ocular allergy. Mast cell localization also changes in the more-chronic forms of ocular allergy, from the substantia propria to the epithelial surface.[20] Interestingly, the histamine concentration in the tears of allergic conjunctivitis patients can reach values greater than 100 ng/mL, as compared with values of 5-15 ng/mL in control patients.[21]

Histamine in the Conjunctiva

In ocular responses, the primary source of histamine is mast cells, which are primarily found below the basement membrane in the substantia propria.[7] Histamine is released in both the early phase and the late phase of allergic reactions. In the early phase response histamine is released from activated mast cells upon degranulation, whereas in the late phase allergic reaction histamine release from mast cells and basophils is dependent on the activity of chemokines and cytokines such as eotaxin, the protein regulated on activation T cell expressed and secreted (RANTES) and interleukins (IL)-1, -3, -5 and -6. These chemokines and cytokines are produced by inflammatory cells such as eosinophils, neutrophils and mononuclear cells and their release can be dependent on or independent of IgE.[22]

Histamine influences the activity of a variety of cell types, including immune cells, vascular cells and epithelial cells. Its actions on immune cells promotes a Th2 response.[23] Histamine is a well-documented vasoactive factor, causing increases in vessel diameter and capillary permeability. In the conjunctiva, vascular responses to histamine increase inflammatory cell recruitment and contribute to the redness and tearing associated with ocular allergy.[24] Finally, histamine can have important effects on the barrier function of the conjunctival epithelial cells.

Histamine and the Conjunctival Barrier

Allergic inflammation predominantly occurs at mucosal surfaces. The epithelium represents a physical barrier that protects against the exaggerated intrusion of antigens through the function of tight junctions and adhesion molecules. Tight junctions are protein complexes that link the cytoskeleton of adjacent epithelial cells thereby increasing adhesion and stability. Tight junction proteins include zonula occludens 1 (ZO-1). One important epithelial adhesion protein is the calcium-dependant adhesion molecule E-cadherin. During allergic inflammation, alterations in expression of these proteins indicate destabilization of the conjunctival epithelium and loss of ocular surface integrity. Lost conjunctival surface integrity could in turn lead to increased penetration of allergen and an exacerbation of the allergic response.

Disruption of the epithelial barrier can occur in response to inflammation or to other factors. Some allergens, including pollen, contain active proteolytic enzymes that break down tight epithelial cell junctions.[25] Other allergens enhance inflammation by stimulating epithelial cell secretion of proinflammatory cytokines and chemokines. Histamine is a particularly potent disruptor of tight junctions. In a study of experimental nasal allergy, locally released histamine caused epithelial cells to decrease ZO-1 mRNA expression and upregulated IL-8 mRNA expression.[26] The former effect may increase the mucosal permeability and penetration of additional antigen through paracellular spaces. The latter effect could stimulate local accumulation of

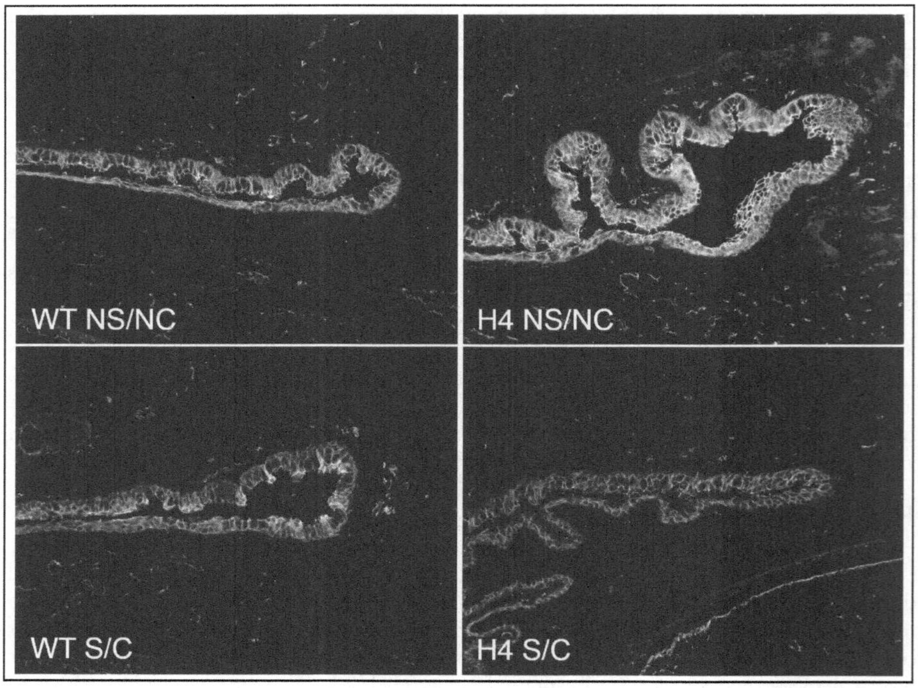

Figure 2. Expression of the epithelial proteins ZO-1 and E-cadherin is similar for nonsensitized/nonchallenged (NS/NC) and sensitized/challenged (S/C) conjunctivae in wild type mice (WT) (left photographs upper and lower), but decreases in expression of these proteins are observed following allergen challenge in H_4R-deficient mice (H4) (right photographs, upper and lower). Green: ZO-1, Red: E-cadherin. A color version of this image is available at www.landesbioscience.com/curie.

eosinophils, thereby enhancing local allergic inflammation. A bronchial biopsy study showed that ZO-1 and E-cadherin were expressed at significantly lower levels in asthmatic subjects than in non-asthmatic subjects.[27] This, too, suggests that disrupted barrier function might contribute to allergic pathogenesis.

Figure 2 shows the expression of the ZO-1 and E-cadherin in the mouse conjunctiva. In the forniceal conjunctiva, ZO-1 is found at the superficial cell layer and E-cadherin is expressed from the basal cell layer to the facial layer. Interestingly, conjunctival expression of these proteins does not decrease following exposure to allergen in a murine model of allergic conjunctivitis (Fig. 2), in contrast to observations for respiratory allergy.[27,28] These differences may be explained by different functions of histamine receptor subtypes.

Histamine Receptors in the Conjunctiva

Histamine receptors are G protein-coupled receptors expressed by a variety of cell types. Of the four known histamine receptors (H_1R-H_4R), the subtypes H_1R, H_2R and H_4R have been most strongly linked to ocular allergy.

Histamine H_1 and H_2 Receptors

H_1 histamine receptors are expressed by a variety of cell types. In the conjunctiva, they are expressed by vascular endothelial cells.[29] The H_1R and H_2R are expressed at higher levels in the conjunctiva of VKC patients compared to normal controls,[29] but expression levels in other types of ocular allergy have not yet been assessed.

Histamine signaling through H_1R and H_2R has been shown to increase conjunctival hyperemia, fibroblast proliferation, cytokine secretion and microvascular permeability.[5] Stimulation of conjunctival H_1R with histamine increases intracellular calcium via inositol phosphate activity, leading to the symptom of pruritus.[5] Histamine stimulation of H_2R on the ocular surface results in vasodilation, another symptom of ocular allergy.[5]

The H_1R plays important roles in both early- and late-phase allergic responses in the eye. H_1R-deficient mice do not display the increased vascular permeability[30] and conjunctival eosinophil infiltration[31] following allergen challenge that are typical in wild-type mice. It is therefore not surprising that this receptor has been targeted by therapeutic agents for ocular allergy. The H_1R antagonist levocabastine reduced vascular permeability[32] and late-phase nitric oxide production[33] in rat models of allergic conjunctivitis. In humans, topical application of levocabastine clinically reduces symptoms of ocular allergy as well as allergic rhinitis;[34] other H_1R antagonists are also used for treatment of ocular allergy.[35] Further effects of histamine on the eye may be due to activity of the newest member of the histamine receptor family, the H_4R.

Histamine H_4 Receptor

The H_4R subtype is expressed on hematopoietic cells such as mast cells, eosinophils, T cells and dendritic cells.[36] The H_4R modulates a variety of physiological functions in these immune cells, including chemotaxis, cytokine and chemokine release and adhesion molecule expression.[36] Conjunctival biopsies from VKC patients display higher levels of H_4R than does the normal conjunctiva, with H_4R especially present in stromal inflammatory cells.[29] Histamine binding to H_4R may therefore selectively recruit mast cells, eosinophils, dendritic cells and T cells into VKC conjunctival tissue.

In vitro, H_4R activation induces mast cell migration towards histamine, though it has no effect on degranulation.[37] The H_4R has also been implicated in the pathogenesis of ocular allergy in vivo. Application of an agonist specific for H_4R resulted in symptoms of ocular allergy in mice.[38] Interestingly, the H_4R antagonist JNJ7777120 was able to inhibit allergic conjunctivitis symptoms induced by both H_1R- and H_4R-specific agonists, whereas selective H_1R antagonists such as levocabastine had no effect on H_4R induced allergic conjunctivitis.[38] These results suggest that H_4R activity is important for allergic conjunctivitis in vivo and that this receptor may even affect H_1R-mediated pathogenesis, though the mechanisms for this remain unknown.

The H_4R may be particularly important in maintaining barrier function of the conjunctival epithelial cells in an allergic response. Although conjunctival epithelial cells in wild-type mice express ZO-1 and E-cadherin at normal levels following allergen challenge, decreases in expression of these tight junction proteins are observed following allergen challenge in H_4R-deficient mice (Fig. 3A). H_4R may therefore maintain the integrity of the conjunctival barrier in the presence of allergen, whereas histamine signaling via H_1R or H_2R may promote disruption of the conjunctival epithelium.

Tissue-Specific Roles of Histamine Receptors

Interestingly, differential effects of H_4R activity have been demonstrated in different tissues. In the lung, H_4R-deficient mice display decreases in inflammation, infiltrating eosinophils and lymphocytes and Th2 responses compared to wild-type mice following exposure to allergen.[39] In a mouse model of allergic conjunctivitis, on the other hand, we have observed comparable eosinophil infiltration between H_4R-deficient mice and wild-type mice (Fig. 3B). Differences in responses to H_1R antagonists also differ depending on the allergy site. Although H_1R antagonists are used to treat a variety of allergic disorders, the H_1R antagonist cetirizine has differing effects when applied prior to allergen challenge in the skin allergy and in allergic rhinitis. Whereas pretreatment with the drug reduces histamine release and migration of eosinophils, neutrophils and basophils at the skin, similar treatment of nasal tissue causes a lesser decrease in mediator release and no change in eosinophil migration.[40] Differential roles of histamine receptors in atopic disease may be influenced by the microenvironment of the tissues involved, a phenomenon that may have important implications for development of therapies for allergic conjunctivitis. Further research is needed to define histamine receptor roles in different tissues during allergic responses.

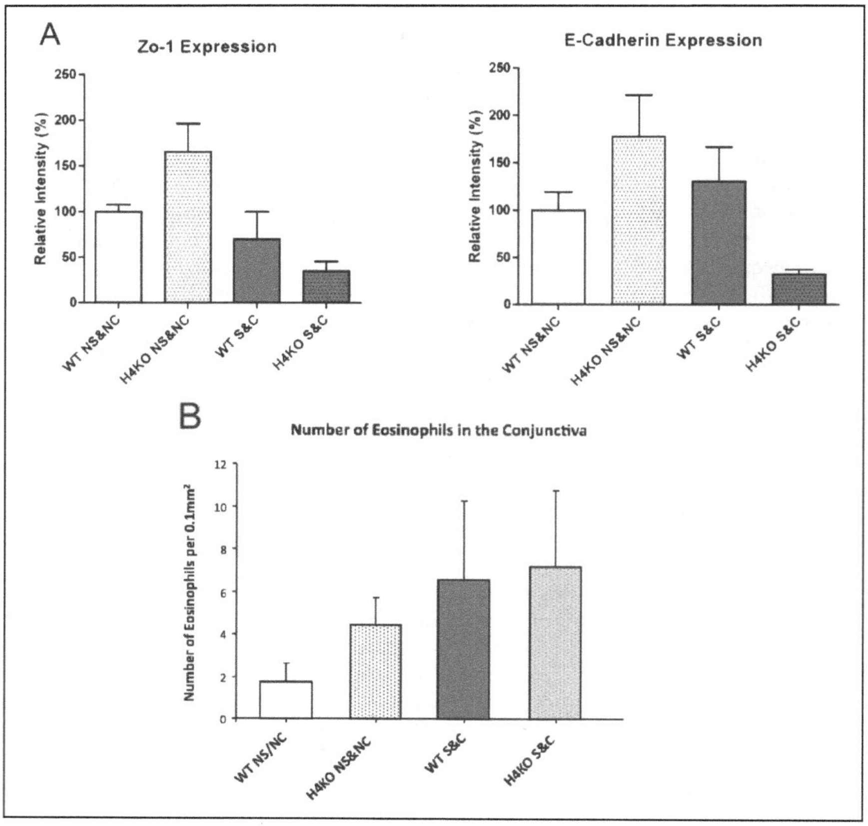

Figure 3. Sensitized challenged H_4R-deficient conjunctivae display decreased expression of ZO-1 and E-cadherin (A), but normal levels of eosinophil recruitment (B). NS/NC, nonsensitized/nonchallenged; S/C, sensitized/challenged; WT, wild type.

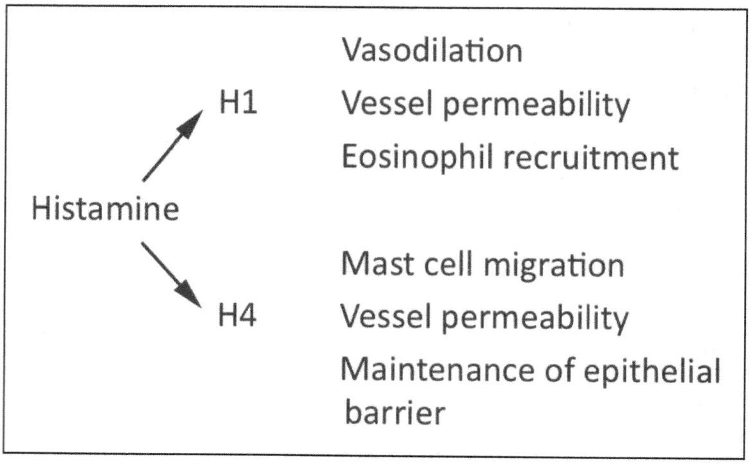

Figure 4. Possible effects of histamine H_1 and H_4 receptors in allergic conjunctivitis.

Conclusion

Histamine clearly plays a critical role in allergic pathogenesis in atopic conditions, including ocular allergy. The exact roles of histamine receptors in different tissues remain poorly understood, but tissue-specific differences in allergic pathogenesis may be mediated by histamine receptor subtypes (Fig. 4). The interplay of histamine with mast cells and other inflammatory cells, such as T cells and eosinophils, is another active area of research with important implications for treatment of allergic conjunctivitis. A better understanding of the pathways and mechanisms involved in allergic diseases at specific sites will greatly assist the development of more-effective pharmacological treatments.

References

1. Stahl JL, Barney NP. Ocular allergic disease. Curr Opin Allergy Clin Immunol 2004; 4(5):455-459.
2. Trocme SD, Sra KK. Spectrum of ocular allergy. Curr Opin Allergy Clin Immunol 2002; 2(5):423-427.
3. Leonardi A. The central role of conjunctival mast cells in the pathogenesis of ocular allergy. Curr Allergy Asthma Rep 2002; 2(4):325-331.
4. Haas HL, Sergeeva OA, Selbach O. Histamine in the nervous system. Physiol Rev 2008; 88(3):1183-1241.
5. Bielory L, Ghafoor S. Histamine receptors and the conjunctiva. Curr Opin Allergy Clin Immunol 2005; 5(5):437-440.
6. Ono SJ, Abelson MB. Allergic conjunctivitis: update on pathophysiology and prospects for future treatment. J Allergy Clin Immunol 2005; 115(1):118-122.
7. Bielory L. Allergic and immunologic disorders of the eye. Part II: ocular allergy. J Allergy Clin Immunol 2000; 106(6):1019-1032.
8. Foster CS, Calonge M. Atopic keratoconjunctivitis. Ophthalmology 1990; 97(8):992-1000.
9. Bonini S, Coassin M, Aronni S et al. Vernal keratoconjunctivitis. Eye 2004; 18(4):345-351.
10. Hoang-Xuan T, Prisant O, Hannouche D et al. Systemic cyclosporine A in severe atopic keratoconjunctivitis. Ophthalmology 1997; 104(8):1300-1305.
11. Akpek EK, Dart JK, Watson S et al. A randomized trial of topical cyclosporin 0.05% in topical steroid-resistant atopic keratoconjunctivitis. Ophthalmology 2004; 111(3):476-482.
12. Rikkers SM, Holland GN, Drayton GE et al. Topical tacrolimus treatment of atopic eyelid disease. Am J Ophthalmol 2003; 135(3):297-302.
13. Vichyanond P, Tantimongkolsuk C, Dumrongkigchaiporn P et al. Vernal keratoconjunctivitis: result of a novel therapy with 0.1% topical ophthalmic FK-506 ointment. J Allergy Clin Immunol 2004; 113(2):355-358.
14. Ormerod AD. Topical tacrolimus and pimecrolimus and the risk of cancer: how much cause for concern? Br J Dermatol 2005; 153(4):701-705.
15. McGill JI, Holgate ST, Church MK et al. Allergic eye disease mechanisms. Br J Ophthalmol 1998; 82(10):1203-1214.
16. Cook EB, Stahl JL, Barney NP et al. Ocular mast cells. Characterization in normal and disease states. Clin Rev Allergy Immunol 2001; 20(2):243-268.
17. Solomon A, Pe'er J, Levi-Schaffer F. Advances in ocular allergy: basic mechanisms, clinical patterns and new therapies. Curr Opin Allergy Clin Immunol 2001; 1(5):477-482.
18. Bonini S, Bonini S. Pathogenesis of allergic conjunctivitis. In: Denburg J, ed. Allergic Diseases: The New Mechanisms and Therapeutics. New York: Humana Press, Inc.; 1998:509.
19. Miyazaki D, Nakamura T, Komatsu N et al. Roles of chemokines in ocular allergy and possible therapeutic strategies. Cornea 2004; 23(8 Suppl):S48-54.
20. Irani AM, Butrus SI, Tabbara KF et al. Human conjunctival mast cells: distribution of MCT and MCTC in vernal conjunctivitis and giant papillary conjunctivitis. J Allergy Clin Immunol 1990; 86(1):34-40.
21. Bielory L. Ocular allergy overview. Immunol Allergy Clin North Am 2008; 28(1):1-23, v.
22. MacDonald SM, Lichtenstein LM. Histamine-releasing factors and heterogeneity of IgE. Springer Semin Immunopathol 1990; 12(4):415-428.
23. Jutel M, Blaser K, Akdis CA. The role of histamine in regulation of immune responses. Chem Immunol Allergy 2006; 91:174-187.
24. Leonardi A. Role of histamine in allergic conjunctivitis. Acta Ophthalmol Scand Suppl 2000; 230:18-21.
25. Runswick S, Mitchell T, Davies P et al. Pollen proteolytic enzymes degrade tight junctions. Respirology 2007; 12(6):834-842.

26. Takeuchi K, Kishioka C, Ishinaga H et al. Histamine alters gene expression in cultured human nasal epithelial cells. J Allergy Clin Immunol 2001; 107(2):310-314.
27. de Boer WI, Sharma HS, Baelemans SM et al. Altered expression of epithelial junctional proteins in atopic asthma: possible role in inflammation. Can J Physiol Pharmacol 2008; 86(3):105-112.
28. Takahashi T, Ono S, Ogawa K et al. A case of anaphylactoid shock occurring immediately after the initiation of second intravenous administration of high-dose immunoglobulin (IVIg) in a patient with Crow-Fukase syndrome. Rinsho Shinkeigaku 2003; 43(6):350-355.
29. De Dominicis C, Brun P, Motterle L et al. Histamine H1R, H2R, H3R and H4R receptors in normal conjunctiva and in vernal keratoconjunctivitis. Invest Ophthalmol Vis Sci 2008; 49:E-Abstract 418.
30. Nakahara H, Izushi K, Sugimoto Y et al. Vascular permeability in allergic conjunctivitis in mice lacking histamine H1 receptors. Eur J Pharmacol 2000; 409(3):313-317.
31. Izushi K, Nakahara H, Tai N et al. The role of histamine H(1) receptors in late-phase reaction of allergic conjunctivitis. Eur J Pharmacol 2002; 440(1):79-82.
32. Bayer A, Uludağ HA, Sobaci G et al. Comparison of antiallergic drugs in an experimental model of ocular anaphylaxis. Ophthalmologica 2003; 217(2):119-123.
33. Papathanassiou M, Giannoulaki V, Tiligada E. Leukotriene antagonists attenuate late phase nitric oxide production during the hypersensitivity response in the conjunctiva. Inflamm Res 2004; 53(8):373-376.
34. Noble S, McTavish D. Levocabastine. An update of its pharmacology, clinical efficacy and tolerability in the topical treatment of allergic rhinitis and conjunctivitis. Drugs 1995; 50(6):1032-1049.
35. Bielory L. Role of antihistamines in ocular allergy. Am J Med 2002; 113 Suppl 9A:34S-37S.
36. Huang JF, Thurmond RL. The new biology of histamine receptors. Curr Allergy Asthma Rep 2008; 8(1):21-27.
37. Hofstra CL, Desai PJ, Thurmond RL et al. Histamine H4 receptor mediates chemotaxis and calcium mobilization of mast cells. J Pharmacol Exp Ther 2003; 305(3):1212-1221.
38. Nakano Y, Takahashi Y, Ono R et al. Role of histamine H(4) receptor in allergic conjunctivitis in mice. Eur J Pharmacol 2009.
39. Dunford PJ, O'Donnell N, Riley JP et al. The histamine H4 receptor mediates allergic airway inflammation by regulating the activation of CD4+ T-cells. J Immunol 2006; 176(11):7062-7070.
40. Simons FE, Simons KJ. The pharmacology and use of H1-receptor-antagonist drugs. N Engl J Med 1994; 330(23):1663-1670.

Chapter 6

The Role of Histamine in Asthma

Paul J. Dunford* and Stephen T. Holgate

Abstract

Histamine is a ubiquitous inflammatory mediator intimately associated with the pathology of allergy. Traditional antihistamines, targeting the histamine H_1 receptor, have failed to demonstrate a significant role for histamine in asthma. Novel immunomodulatory roles for histamine and the discovery of a novel histamine receptor, the histamine H_4 receptor, have resulted in a reassessment of its importance in asthma.

Introduction

Asthma is a complex and increasingly prevalent airway syndrome, most often associated with an allergic phenotype. It is characterized by inflammatory infiltrates and episodes of reversible airway obstruction. In its more serious forms chronic airway remodeling may result in persistent airway dysfunction and hyperreactivity.

The association of asthma with allergy has long been apparent, with the atopic phenotype of immunoglobulin-E, mast cell, eosinophil and T-helper-type 2 (Th2) cytokine involvement, commonly displayed. Correspondingly, IgE-mediated mast cell degranulation is thought to be involved in the pathogenesis of allergic asthma through the resultant release of potent physiological and immunological modulators such as eicosanoids, proteases and histamine.

Histamine exerts its effects in the airways, potentially through all four histamine receptors. A pathological role for histamine in asthma has long been proposed, largely due to its increased presence in asthmatic patients and its ability to manifest many symptoms of asthma when administered to the human lung. In spite of this association, frequent and robust analysis of drugs designed to block the action of histamine at the histamine H_1 receptor, namely 'antihistamines', have failed to demonstrate meaningful clinical benefit on symptoms or progression of allergic asthma disease.

More recently a role for histamine in immune modulation has been proposed via its activity at various histamine receptors, notably the histamine H_1 (H_1R), H_2 (H_2R) and H_4 (H_4R) Whilst relatively high concentrations of histamine, as detected in asthmatic airways might be needed to activate low affinity H_1R and H_2R, it is ironic, given the previously stated correlates between high histamine levels and asthma, that minute levels of histamine, acting at high affinity receptors such as H_4R may be more pathologically relevant to the immune malfunction observed in asthma development and persistence. These traditional and emerging hypotheses for the role of histamine in asthma are discussed below.

Histamine in the Asthmatic Airway

Histamine has been considered to be intimately associated with the pathophysiology of asthma since its identification as a potent constrictor of airway smooth muscle by Henry Dale in 1910[1] and the subsequent realisation of its increased presence in diseased tissue by Curry in 1946[2] In

*Corresponding Author: Paul J. Dunford—Johnson & Johnson PRD, LLC, Immunology, 3210 Merryfield Row, San Diego, California 92121, USA. Email: pdunford@its.jnj.com

Histamine in Inflammation, edited by Robin L. Thurmond.

addition, the main source of histamine in the airway, IgE stimulated mast cells, have been shown to be increased in the airways of allergic asthmatics.[3,4] Their infiltration into airway smooth muscle in asthmatics has also been associated with airway dysfunction.[5] Basophils, were also demonstrated to be a relevant source of histamine in anti-IgE stimulated blood[6] and have been noted in increased numbers in the BAL fluid of allergen challenged asthmatics and in bronchial biopsies of allergic and non-allergic asthmatics,[7] particularly during the late phase response, as has also been noted with mast cells.[4]

Correspondingly, upon antigen challenge of the airways, histamine levels are rapidly increased in the airways and in plasma.[8] This increase has been correlated with disease severity,[3,9-13] whilst effective management of asthma lowers plasma histamine.[14] Since the pharmacological actions of histamine on the airways and cells involved in the asthmatic response mimic many of the pathophysiological features of asthma, these observations are additionally provocative in terms of a pathologically relevant role for histamine in the disease.

Physiologic Role for Histamine in Lung and Asthma

One of the first identified actions of histamine was its constrictor actions on airway smooth muscle.[1] This response was subsequently shown to exacerbate disease when Weiss reported that administration of histamine to asthmatic patients resulted in breathlessness and decreased vital capacity.[15] This association was strengthened with the observation that the bronchoconstrictor response to histamine was enhanced in asthmatic versus normal individuals and that these responses could be blocked by a prototypical H_1R antihistamine drug.[2] Subsequently, development of more selective H_1R antagonists allowed for studies to more clearly define the role of histamine and H_1R in bronchoconstriction. In one important study, inhalation of histamine leading to bronchoconstriction and a decrease in FEV_1 (forced expiratory volume in 1 second), was significantly inhibited by oral administration of all H_1R antagonists tested, whose efficacy in the lung correlated well with that observed against histamine induced skin responses in the same patients. In addition, none of the antihistamines were able to inhibit methacholine induced bronchoconstriction.[16] Consequently, this one study demonstrated the selectivity and specificity of the H_1R mediated bronchoconstrictor response to histamine, whilst supporting the theory that the histamine induced vasodilatory responses in skin and bronchoconstrictor responses in lung are mediated by the same receptor.

While histamine is incontrovertibly a constrictor of large and small airway smooth muscle, the direct contractile effect of histamine on airway smooth muscle cells is largely inferred from in vitro studies on human tissue. There is some debate as to whether, in vivo, H_1R present on sensory nerves may contribute to an indirect contractile response via stimulation of a vagally mediated parasympathetic reflex. Whilst this may be relevant in some species, the lack of effect of anticholinergics on histamine-induced bronchoconstriction in humans would seem to argue against this. In addition, a role for histamine in maintaining bronchial smooth muscle tone, in the absence of nerve innervation has also been demonstrated.[17] Constitutive production of histamine and cysteinyl leukotrienes appeared to impart a resting tone on bronchial smooth muscle that could be significantly reversed by addition of H_1R antagonists and CysLT1 receptor antagonists, respectively. Provocatively, H_1R antagonists dosed to asthma patients can produce immediate bronchodilatory responses, with one study demonstrating that cetirizine was almost as potent a bronchodilatory agent as a β-adrenergic agonist.[18] However, as discussed in detail below, these effects are inconsistent and appear to fade with continuous treatment.

The previously mentioned edematous and vasodilator responses observed in the skin to histamine could additionally have pathological implications in asthmatic airways via the causation of mucosal edema and the facilitation of plasma proteins and leukocyte movement into the affected tissue. These vasodilator effects are mediated by H_1R on vascular endothelium which act to increase paracellular permeability,[19,20] whilst the movement of cells is additionally facilitated by the H_1R dependent upregulation of adhesion molecules such as ICAM-1, E-selectin and P-selectin on endothelial and epithelial cells,[21-23] particularly under inflammatory conditions. Histamine acting on H_1R on endothelium and airway epithelium may have direct pro-inflammatory effects via the release of cytokines such as IL-6 and IL-8[24,25] and has also been shown to augment the release of IL-16, a potent chemokine for T helper cells, from airway epithelial cells.[26] Interestingly, histamine has been shown to similarly induce IL-16 release from $CD8^+$ T cells in a H_4R and H_2R dependent fashion.[27]

All of these actions of histamine on airway structural cells may combine to promote an inflammatory phenotype, however a caveat of all these data is that they have largely been obtained in vitro and their in vivo significance is unknown. Likewise, a role for histamine in airway smooth muscle cell proliferation in vitro has also been reported[28] and other studies have suggested histamine receptors to be involved in the hypersecretory response seen in asthmatic airways. Whilst H_1R may be linked to airway secretion via an effect secondary to its role in plasma exudation, the H_2R has been shown to be the sole histamine receptor involved in mucus secretion,[29] which is consistent with the expression of H_2R in secretory cells from nasal mucosa.[30]

In summary, the pharmacological effects of histamine on the lung recapitulate many of the pathophysiological symptoms of asthma and that the potential role of histamine as an important mediator in asthma is further suggested by its increased presence in the asthmatic airway. In the rest of this chapter we will examine whether this 'guilt by association' is borne out by the evidence from clinical studies with antihistamines and whether newer insight into the role of histamine in immune modulation, potentially through receptors not targeted by current antihistamines, may have implications for the future treatment of asthma.

Immunological Modulation by Histamine

Many cells of the immune system associated with asthma have been shown to express a range of histamine receptors, notably H_1R, H_2R and the newly described H_4R. In most cases these receptors are co-expressed so that the net effect of their activation may vary, depending on the exact expression profile and the concentration of histamine in the surrounding milieu, and is further complicated since the receptors may have opposing function. It is therefore difficult to interpret the significance of the in vitro literature in which many of these conditions may be manipulated to observe a specific effect and hence preclinical examination of their role in the whole animal is likely the most relevant predictor of their role in human disease. Nevertheless, the in vitro literature reviewed here demonstrates the broad spectrum of cells and effector functions that may be modulated by histamine. In vivo evidence will be described later. The proposed immune and physiologic roles of histamine in asthma are depicted in Figure 1.

Mast Cells and Basophils

Mast cells are not only the main source of histamine in the lung, but are a source of cytokines and tissue growth factors that may be important in the inflammatory and remodeling processes observed in asthma.[31] They themselves may be modulated by histamine through expression of H_1R, H_2R and H_4R on their surface. While histamine does not appear to affect degranulation of mast cells, it has been shown to be a potent chemoattractant for mast cells[32] and has been shown to enhance chemotaxis of mast cell precursors in response to CXCL12,[33] all via activity at the H_4R. In addition, inhaled histamine was able to increase the number of sub and intra-epithelial mast cells in mouse airways in an H_4R dependent fashion.[34] Interestingly, localization of mast cells to the bronchial epithelium has been observed in asthmatics after allergen challenge.[35] There are several reports of H_1R antagonists reducing leukotriene and histamine release from human lung mast cells and basophils , but these effects are believed to be independent of their antagonism of histamine, as previously reviewed.[36] Histamine acting at the H_2R does appear to have inhibitory effects on mast cells, decreasing histamine and cytokine release[37,38] and may have similar actions on basophils.[37] H_4R has also been detected on basophils but its role in their function has yet to be studied.

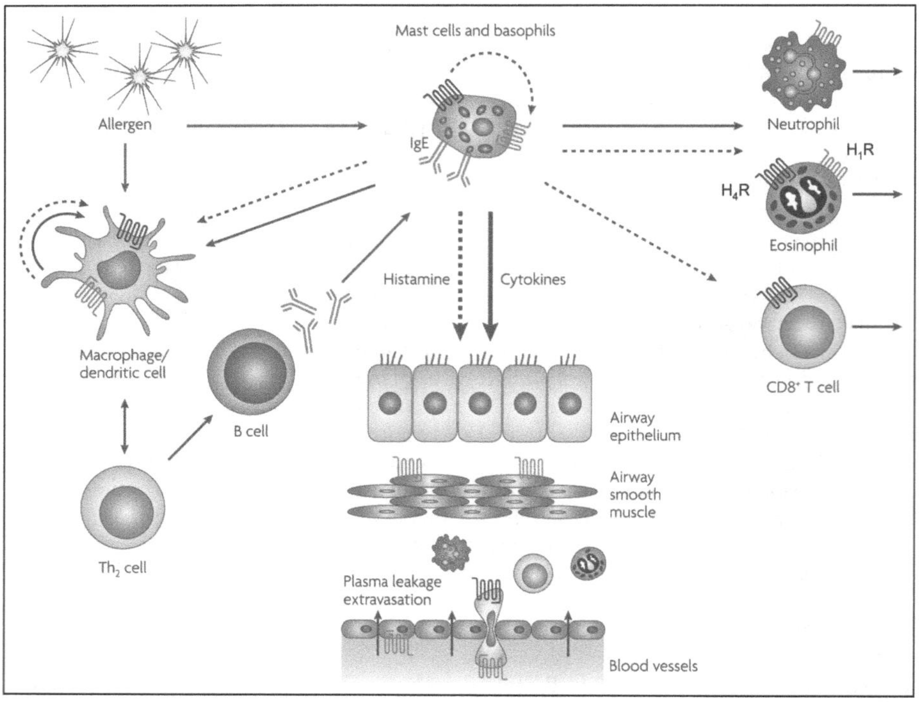

Figure 1. Potential role of histamine in asthma. Allergen entering the airways may cross-link IgE on mast cells(or basophils) to release histamine, lipid mediators and cytokines. Antigen is also processed by airway dendritic cells and macrophages for presentation to T helper cells. During this process local release of histamine and cytokines may also occur. Resultant histamine from these processes can act at a variety of cells and levels. Histamine can facilitate the recruitment of inflammatory cells via direct chemotaxis of additional dendritic cells, eosinophils and mast cells to the airways via action at H_4R, whilst also aiding the chemotactic and inflammatory process through effects at H_1R on the airway epithelium and vascular endothelium. Release of cytokines, such as IL-8, from the airway epithelium and increased vascular permeability in response to H_1R activation enrich the inflammatory milieu. Constriction and proliferation of airway smooth muscle via H_1R also contributes to the asthma phenotype. Histamine has diverse effects on the activation of leukocytes via H_1R and H_4R. Complex autocrine and paracrine processes in response to dendritic cell histamine and cytokine release control the priming and education of T cells via cytokines that are released from dendritic cells in response to H_1R and H_4R ligation, during antigen presentation. Histamine may additionally affect the cytokine release from CD8+ cells via H_4R and from mast cells, eosinophils and neutrophils through multiple histamine receptor activity. Key: solid arrows: cytokine effect; dashed arrow: histamine effect; light 7TM: H_1R; dark 7TM: H_4R. A color version of this image is available at www.landesbioscience.com/curie. Reproduced from Thurmond et al. The role of Histamine H1 and H4 receptors in allergic inflammation: the search for new antihistamines. Nat Rev Drug Discovery 2008; 7:41-53.

Eosinophils

Eosinophils are traditionally one of the major cell types implicated in the pathology of asthma, due to their observed increase in asthmatic lungs and the plethora of cytotoxic and pro-inflammatory mediators, linked with disease progression, that they are able to release. Although a direct causative role in disease has been difficult to prove, recent pharmacologic intervention has suggested their importance in at least a subset of asthma patients.[39,40] Therefore, it is provocative to note that, as for mast cells, histamine has been shown to be a potent chemoattractant and enhancer of chemokine mediated chemotaxis, once more through the recently described H_4R.[41,42] H_4R dependent upregulation of adhesion molecules was also reported.[42] H_1R antagonists have been investigated for their role in eosinophil function, but effects via nonhistaminergic pathways at irrelevant concentrations confuse the interpretation of these studies. As with mast cells and basophils, these activities appear independent of H_1R activity and are not discussed here, as they do not contribute to an understanding of the role of histamine in asthma.

Neutrophils

The presence of neutrophils in the airways has been demonstrated during asthma exacerbations, after treatment withdrawal[43] and in a sub-set of chronic asthma patients with severe disease.[44,45] In patients that die suddenly from asthma, they are also increased,[46] with the same study demonstrating elevated histamine levels. Interestingly, neutrophils have also recently been identified as a source of histamine in the lung,[47] so this correlation may be effect rather than cause. The role of histamine in neutrophil function is also unclear, with any function appearing limited to the H_2R, which may in fact have a negative regulatory role on their function.[48] Histamine may have an indirect effect on neutrophil chemotaxis via its H_1R-dependent ability to stimulate release of proneutrophilic cytokines and chemokines from airway tissue.[24,25] Similarly, an indirect role for H_4R has been reported in mast cell dependent models of neutrophilia.[42,49,50]

Monocytes and Macrophages

Expression of the H_1R, H_2R and H_4R have been demonstrated in human monocytes yet their expression on different macrophage lineages and at different levels of activation may vary and has not been well studied. Alveolar macrophages are the most abundant inflammatory cell in the human lung, yet their association with asthma is not clear, likely due to the difficulty in studying this highly heterogenous population. However, alveolar macrophage suppression of T-cell proliferation is reduced in asthma and after allergen challenge[51] and some alterations in specific subpopulations have recently been described in asthma.[52] In vitro studies have demonstrated that histamine may have modulatory effects on LPS stimulated monocytes via the H_2R, including a reduction in the production of IL-12 and an increase in IL-10 release,[53,54] which could conceivably promote Th2 cell development. The constitutive production of MCP-1 (CCL2) has also been reported to be inhibited via an action at the H_4R.[55] In alveolar macrophages, specifically, H_1R mediates histamine induced β-glucuronidase and IL-6 release, which may indicate a role for histamine in macrophage-mediated remodeling processes.[56]

Dendritic Cells

Dendritic cells are professional antigen presenting cells that may develop from cells of either lymphoid or monocytic lineage. They are intimately associated with the pathogenesis of asthma through their initiation and maintenance of T-cell responses, particularly Th2 type. Polarization of naïve Th0 cells to Th2 and other T helper sub-sets may be differentially controlled at the level of the interaction between dendritic cells and antigen-specific T cells. Such interaction can be directed by a variety of cytokines, chemokines, toll-ligands and biogenic amines, such as histamine. These are released at sites where antigen is encountered or presented and may sequentially modulate both the dendritic cell and subsequent T helper phenotypes.[57] All four histamine receptors have been identified on immature and mature dendritic cells. Histamine, released from dendritic cells or more traditional sources, may act in an autocrine or paracrine fashion to modify their phenotype, as measured by alterations in surface markers[58], or in cytokine release.[59] Cytokine secretion, including inhibition of IL-12 and enhancement of IL-10 and IL-6, may be modulated by histamine with H_1R, H_2R and H_4R all involved[60-62] in these Th2 promoting processes. The autocrine activation of dendritic cells by histamine deserves additional discussion, since it is indicative that low levels of locally released histamine are able to define and control immune responses that may be important in asthma, via actions at high affinity histamine receptors, such as H_4R and at concentrations where low affinity H_1R and H_2R may not be engaged. This may suggest that the high levels of histamine frequently cited as correlating with asthma severity are unrelated to the underlying immunology of asthma and therefore, at best, may only have a relationship to the physiological *sequalae* previously described.

T Cells

As described above, T cells are pivotal cells in the initiation and perpetuation of adaptive immune response associated with allergic asthma. In addition to being modulated by the effects of histamine on dendritic cells, they may be directly affected by histamine. H_1R, H_2R and H_4R are all expressed on CD4+ and CD8+ cells and have been shown to demonstrate reciprocal responses to histamine, based on the preferential expression of H_1R on Th1 cells and H_2R on Th2 cells, for example.[63] Most recently, H_4R has also been described to be functionally active on human Th2 cells[64], with upregulation of H_4R in response to IL-4 reported. H_4R agonism of these cells resulted in activation of the pro-Th2 transcription factor, AP-1 and the induction of the Th2 cytokine, IL-31. H_1R on Th1 cells appears to enhance Th1 type responses, with deletion in mice leading to a consequent skewing to Th2 type responses after T-cell dependent antigen immunization and resultant enhanced production of IgE and IgG_1.[65] H_2R appears to negatively regulate both Th1 and Th2 responses, with the surprising net effect of H_2R deletion in mice resulting in decreased IgE in response to immunization, at least in the Th2 predominating system tested.[63] This was in spite of the predicted increase in IL-4 and IL-13 production from H_2R deficient mice, suggesting that the concomitant overproduction of IFNγ had a dominant effect on the humoral response.

In vitro, T-cell proliferation has also been reported to be affected by histamine, with once more a pro- inflammatory enhancement of proliferation associated with H_1R activation[63,66]and an inhibitory effect on proliferation via H_2R activation, reported.[63,67] On CD8+ T cells deletion of either H_1R or H_2R has been shown to increase their capacity for IFN-γ release whilst reducing IL-2 and IL-10 secretion.[68] Activation of H_2R and H_4R on CD8+ cells also leads to IL-16 release, a potent T-cell cheomattractant associated with asthma.[27] Direct effects of histamine on the chemotaxis of T cells are also apparent, via activity at either H_1R or H_4R.[69,70] The effect of histamine on T-cell and dendritic cell interactions, through action at H_4R is depicted in Figure 2.

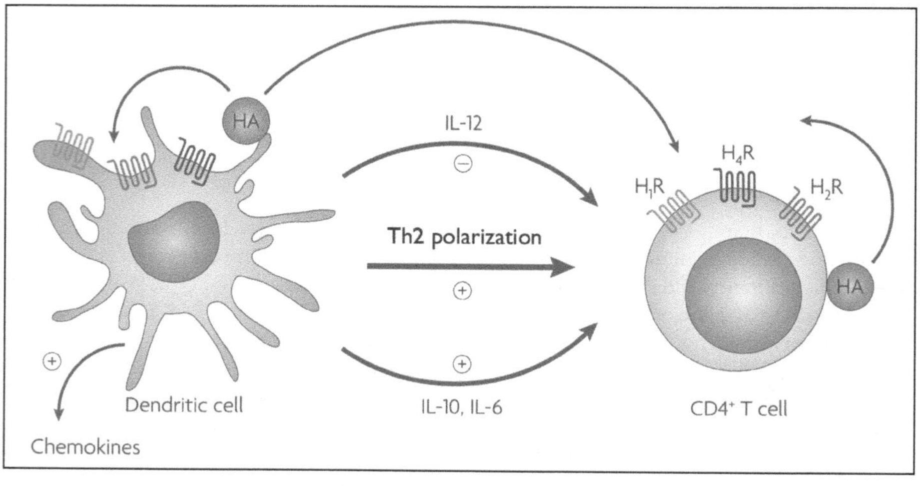

Figure 2. Role of histamine H_4R on dendritic cell and T-cell function. Histamine acting at H_4R on dendritic cells can drive T helper 2 (Th2) cell polarization. This may be accomplished by inhibiting Th1 polarizing cytokines such as IL-12 and by stimulating cytokines such as IL-10 and IL-6. Release of chemokines such as TARC, MDC and MIP-1α may further influence this and the further recruitment of Th2 cells to sites of antigen capture. Histamine may also act as a direct chemoattractant at H_4R on Th2 cells. Reproduced from Thurmond et al. The role of Histamine H1 and H4 receptors in allergic inflammation: the search for new antihistamines. Nat Rev Drug Discovery 2008; 7:41-53.

iNKT Cells

Although still somewhat controversial, invariant NKT cell are considered to be a potentially important initiator of allergic and inflammatory responses, via their ability to rapidly produce primary cytokines such as IL-4 and IFN-γ.[71] Allergens such as pollen may also act at the invariant TCR to cause their activation, further linking them to allergic conditions.[72] Modulation of iNKT cells by histamine has recently been reported, whereby histamine deficient mice demonstrated reduced IL-4 and IFN-γ production in response to in vivo iNKT activation. This could be reconstituted with histamine and blocked by a selective H_4R antagonist.[73]

In concluding this section, the diverse roles of histamine in asthma associated immune cells, dissected from in vitro experiments, would appear to argue for a potential immunomodulatory role for histamine in asthma. The next sections, however, will highlight the caveats of in vitro histamine research, discussed earlier and review the contribution of histamine to in vivo models of asthma and human disease, based on genetic, pharmacological and clinical evidence.

Histamine in Animal Models of Asthma

A role for histamine in the pathogenesis of asthma has also been studied using animal models of asthma. Notably, the role of H_1R has been extensively studied with genetic knockout and pharmacological approaches, while that of H_4R has just begun. Interpretation of these studies is complicated by the exact nature of the model and the dosing regimen applied to each investigation. In general, models have a 'sensitization' phase to allergen, which is comparable to the development of allergy in patients and a 'challenge' phase which is considered analogous to the pro-inflammatory response to an allergen in an already allergic individual, such as an atopic asthmatic. Hence pharmacological intervention around the challenge phase is considered most translational to the therapeutic treatment of pre-existing asthma in patients.

Correspondingly, when H_1R antagonists have been examined therapeutically in mice, and at clinically relevant doses, they appear to have no effect on allergic airway inflammation.[62,74] These respective studies used loratadine and desloratadine and the results are therefore consistent, being parent and metabolite, respectively. Others have tested high doses of alternate H_1R blockers, such as fexofenadine and have demonstrated an effect with a similar dosing schedule in such models.[75] Of interest, higher than standard doses of H_1R have also been shown to have more efficacy in urticaria and are the current recommended treatment.[76] However, these effects may not results completely from blockade of H_1R and as with non-H_1R mediated effects of antihistamines in vitro these findings do not contribute to this discussion of the role of histamine in asthma. Interestingly, high dose loratadine was ineffective in the study previously referenced,[62] further supporting the hypothesis that the reported effects of fexofenadine may be attributed to non-H_1R activity of the drug.

In contrast, H_1R blockade around the sensitization phase of these models appears to indicate a role for H_1R in the developmental stage of allergic responses. In one study, clinically relevant doses of desloratadine were administered prior to antigen sensitization , resulting in a reduced inflammatory response to allergen challenge.[74] Specifically, eosinophilic and lymphocytic airway inflammation was reduced with a resultant decrease in airway hyperreactivity in treated animals. Th2 cytokines and immunoglobulins were also decreased, confirming the inhibitory effect of desloratadine on immune sensitization. The effect of H_1R blockade on allergic sensitization may also explain the reported protective effect of H_1R genetic deficiency in mouse asthma models.[69,77] H_1R deficient mice had a diminished airway inflammation and Th2 cytokine response to allergen challenge as well as a decrease in airway hyperreactivity.[69] Since in these animals the H_1R is absent at all stages of the model the exact point of effect is difficult to pinpoint, but in these studies, transferring cells from sensitized H_1R deficient mice into wild-types which were then challenged with allergen recapitulated the effect. This reinforces the pharmacological evidence that it is an effect at sensitization that ameliorates the subsequent response to allergen challenge. There is also evidence of synergy between the H_1R and H_4R during the sensitization phase.[78] Some clinical studies suggest that these findings may be relevant to humans, since treatment of atopic children with antihistamines appeared to reduce the risk of developing asthma in later life.[79]

Other histamine receptors have been studied in similar models. H_2R genetic deficiency resulted in a decrease in ova-specific IgE production which was attributed to an apparent increase in IFN-γ production in these mice.[65] Pharmacological administration of H_2R antagonist prior to challenge did not affect IgE levels, but did diminish eosinophilia and airway hyperresponsiveness,[80] whilst low dose H_1R antagonist was once more ineffective. However, the effect of H_2R antagonist in this model do not mirror the non-effect of H_2R antagonists observed in clinical asthma,[81,82] perhaps highlighting the difficulty in interpreting animal data.

Despite this, compelling in vivo evidence does exist for a role of the newly described H_4R in asthma. In a mouse ovalbumin challenge model, H_4R deficient animals were protected from eosinophilic and lymphocytic airway inflammation with concomitant reduction in Th2 cytokines in airway and draining lymph nodes. Ova-specific IgE was also decreased.[62] In addition, pharmacological intervention at either the sensitization or challenge stage lead to a diminishment of the subsequent allergic response, suggesting H_4R is involved in both the developmental and effector phases of allergic inflammation. Dissection of the cellular basis for these effects indicated that inhibition of H_4R on dendritic cells results in a decrease in their capacity to activate Th2 cells.

Taken together, these investigations demonstrate the ability of histamine receptors to contribute to various stages of allergic inflammation in mouse models. The therapeutic translation of these findings is discussed below.

Antihistamines and Clinical Asthma

This weight of preclinical evidence and the provocative similarity between the physiological effects of histamine and the pathological symptoms of asthma, has resulted in an enduring investigation into the possible utility of antihistamines in asthma, over the past 50 years.

Promising results were initially obtained with first generation antihistamines demonstrating their acute bronchodilator effects in asthmatic patients[83] and against histamine and allergen induced bronchoconstriction, that appeared independent of muscarinic activity.[84] Despite this, doses that were absent of dose-limiting sedation were ineffective in modulating exercise or naturally occurring asthma,[81] whilst inhaled antihistamines, aimed at reducing these side-effects, were either irritants or variable in their observed activity.[85-88] Consequently, it was not until the development of more potent and nonsedative second generation antihistamines in the 1980s that the role of antihistamines in asthma was reappraised. A selection of these studies is summarized in Table 1.

Whilst a meta-analysis of properly designed studies reached an overall conclusion that second generation antihistamines were ineffective in the treatment of asthma,[89] some interesting trends are apparent that require further discussion. Notably, doses of antihistamines much greater than those observed to be effective in rhinitis appear to offer some benefit and concomitant therapy with other agents might also be beneficial. Several studies with recommended doses of terfenadine, for example, while demonstrating an ability to produce bronchodilation,[16] or reduce early phase responses to allergen challenge,[90] showed no effect on late phase responses.[91] However, increasing doses of terfenadine were able to enhance the effect on early phase responses and afford protection against late phase responses.[92] A similar reliance on increased doses of terfenadine for efficacy in clinical asthma has also been observed. In comparable double-blind placebo controlled studies in mild asthmatics, terfenadine was more active than placebo when dosed at 120 mg twice daily,[93] or at 180 mg thrice daily.[94] The higher dose appeared more effective at increasing peak expiratory flow and reducing use of inhaled beta-agonists, as well as reducing cough and wheeze. A suggestion that disease severity may also determine the efficacy of antihistamines was suggested in another trial where 120 mg twice a day terfenadine was deemed inactive in a population of severe asthmatics.[95] A more recent low dose trial of fexofenadine, the active metabolite of terfenadine, in moderate disease, demonstrated a mild but transient effect on wheezing and peak expiratory flow.[96] This transient effect may be related to a loss of the bronchodilator effect that has been observed with repeat dosing of antihistamines, including terfenadine.[97]

Similar trends to efficacy have been seen with other antihistamines such as cetirizine and loratadine. Cetirizine has demonstrated improvement in lung function and symptoms over placebo,[98] while

Table 1. Clinical assessment of second and third generation antihistamines in asthma

Antihistamine (Ref)	Dose	Asthma Phenotype	Comparator/ Additional Drug (+)	Duration of Study	Symptoms	Function	β-Agonist Use
Cetirizine[98]	10 mg bid	Seasonal	Placebo	2 wk	Decreased	Improved	Decreased
Cetirizine[103]	10/15 mg bid	Seasonal	Terfenadine 60 mg bid	4 wk	Decreased (15 mg only)	No change	No change
Fexofenadine[96]	180 mg qd	Moderate, seasonal		6 wk	Decreased wheezing	Improved PEF	No change
Loratadine[101]	20 mg qd	Moderate to severe, perennial		4 wk	No change	Transient improvement	No change
Loratadine[99]	5 mg bid	Mild, seasonal	+ pseudoephedrine 120 mg bid	6 wk	Decreased	Improved PEF	No change
Loratadine[100]	20 mg qd	Perennial	+ Montelukast 10 mg qd	2 wk	Decreased	Improved	Decreased
Terfenadine[93]	120 mg qd	Mild, perennial		2 wk	Decreased	Improved PEF	Decreased
Terfenadine[92]	180 mg tid	Mild, seasonal		4 wk	Decreased	Improved PEF	Decreased (NS)
Terfenadine[104]	120 mg bid	Severe, perennial		4 wk	Decreased night time awakening	No change	No change

Reproduced from Nelson HS. Prospects for antihistamines in the treatment of asthma. J Allergy Clin Immunol 2003; 112:S96-S100; ©2003 with permission from Elsevier.

another study showed a dose dependent improvement in symptoms, in comparison to terfenadine, but no effect on lung function.[97] Once more, relatively high doses with some sedative effects were used. The beneficial effects with loratadine have most interestingly been shown in combination with other drugs. The combination of loratadine with pseudoephedrine has been reported to improve both rhinitis and asthma symptoms, with an improvement in peak expiratory flow measurements, versus placebo.[99] No comparison to individual agents alone was made, however. The addition of loratadine to montelukast therapy was examined by Reicin and colleagues.[100] The dual treatment provided a significant improvement in asthma symptoms and lung function, as determined by peak expiratory flow and FEV_1, when compared to montelukast alone. Mild but transient improvement in lung function has also been observed with high dose loratadine alone,[101] perhaps reiterating the apparent development of toleration to the bronchodialtor effects of antihistamines.

In summary, the limited efficacy of H_1R antagonists in asthma does not appear to warrant their use over existing steroid sparing therapies, such as montelukast, although cotherapy may be warranted in some instances. The use of totally nonsedating blockers such as fexofenadine at higher doses may prove fruitful in the future. Whether this apparent improved efficacy at high doses is a result of additional ant-inflammatory effects is not clear, but the studies in mice discussed earlier are beginning to tease out the possible immunomodulatory role of H_1R antagonists[102] and may additionally explain the observed prophylactic effect of antihistamines in reducing asthma development in atopic children.[79]

Conclusion

Whilst there is a plethora of preclinical data supporting a role for histamine in the pathophysiology of asthma, the general lack of efficacy of H_1R antagonists in the clinical setting is perplexing and may speak to the importance of histamine's effects on other cell types, mediated by other receptors. Activation of the H_2R certainly mimics some pathophysiological aspects of asthma, but H_2R antagonists have no efficacy in allergic asthma. Therefore, the role of histamine in the modulation of immune cell chemotaxis, function and education, with the recent observations of an important role for H_4R in these events, may provide an additional and novel avenue for successful pharmacological modulation of histamine mediated responses. In addition, to this novel receptor pharmacology some positive studies investigating higher doses of second generation H_1R antagonists in allergic asthma, may help reconcile the current discrepancy between the observed experimental importance of histamine and its apparent clinical irrelevance.

References

1. Dale HH, Laidlaw PP. The phsiological action of beta-iminazolylethylamine. J Physiol 1910; 41:318-344.
2. Curry JJ. The effect of antihistamine substances and other drugs on histamine bronchoconstriction in asthmatic subjects. J Clin Invest 1946; 25(6):792-799.
3. Wardlaw AJ, Dunnette S, Gleich GJ et al. Eosinophils and mast cells in bronchoalveolar lavage in subjects with mild asthma. Relationship to bronchial hyperreactivity. Am Rev Respir Dis 1988; 137(1):62-69.
4. Crimi E, Chiaramondia M, Milanese M et al. Increased numbers of mast cells in bronchial mucosa after the late-phase asthmatic response to allergen. Am Rev Respir Dis 1991; 144(6):1282-1286.
5. Brightling CE, Bradding P, Symon FA et al. Mast-cell infiltration of airway smooth muscle in asthma. N Engl J Med 2002; 346(22):1699-1705.
6. Ishizaka K, Ishizaka T, Terry WD. Antigenic structure of gamma-E-globulin and reaginic antibody. J Immunol 1967; 99:849-858.
7. He S, Gaca MDA, McEuen AR et al. Inhibitors of chymase as mast cell-stabilizing agents: contribution of chymase in the activation of human mast cells. J Pharmacol Exp Ther 1999; 291(2):517-523.
8. Busse WW, Swenson CA. The relationship between plasma histamine concentrations and bronchial obstruction to antigen challenge in allergic rhinitis. J Allergy Clin Immunol 1989; 84(5, Pt 1):658-666.
9. Broide DH, Gleich GJ, Cuomo AJ et al. Evidence of ongoing mast cell and eosinophil degranulation in symptomatic asthma airway. J Allergy Clin Immunol 1991; 88(4):637-648.
10. Casale TB, Wood D, Richerson HB et al. Elevated bronchoalveolar lavage fluid histamine levels in allergic asthmatics are associated with methacholine bronchial hyperresponsiveness. J Clin Invest 1987; 79(4):1197-1203.

11. Jarjour N, Calhoun W, Schwartz L et al. Elevated bronchoalveolar lavage fluid histamine levels in allergic asthmatics are associated with increased airway obstruction. Am Rev Respir Dis 1991; 144(1):83-87.
12. Liu MC, Bleecker ER, Lichtenstein LM et al. Evidence for elevated levels of histamine, prostaglandin D2 and other bronchoconstricting prostaglandins in the airways of subjects with mild asthma. Am Rev Respir Dis 1990; 142(1):126-132.
13. Wenzel SE, Fowler AA 3rd, Schwartz LB. Activation of pulmonary mast cells by bronchoalveolar allergen challenge. In vivo release of histamine and tryptase in atopic subjects with and without asthma. Am Rev Respir Dis 1988; 137(5):1002-1008.
14. Akagi K, Townley RG. Spontaneous histamine release and histamine content in normal subjects and subjects with asthma. J Allergy Clin Immunol 1989; 83(4):742-749.
15. Weiss S, Robb GP, Ellis LB. The systemic effects of histamine in man. Arch Intern Med 1932; 49:360-396.
16. Wood-Baker R, Holgate ST. The comparative actions and adverse effect profile of single doses of H1-receptor antihistamines in the airways and skin of subjects with asthma. J Allergy Clin Immunol 1993; 91(5):1005-1014.
17. Ellis JL, Undem BJ. Role of cysteinyl-leukotrienes and histamine in mediating intrinsic tone in isolated human bronchi. Am J Respir Crit Care Med 1994; 149(1):118-122.
18. Spector SL, Nicodemus CF, Corren J et al. Comparison of the bronchodilatory effects of cetirizine, albuterol and both together versus placebo in patients with mild-to-moderate asthma. The J Allergy Clin Immunol 1995; 96(2):174-181.
19. Rotrosen D, Gallin JI. Histamine type I receptor occupancy increases endothelial cytosolic calcium, reduces F-actin and promotes albumin diffusion across cultured endothelial monolayers. J Cell Biol 1986; 103(6, Pt 1):2379-2387.
20. Ehringer WD, Edwards MJ, Miller FN. Mechanisms of a-thrombin, histamine and bradykinin induced endothelial permeability. J Cell Physiol 1996; 167(3):562-569.
21. Kubes P, Kanwar S. Histamine induces leukocyte rolling in postcapillary venules. A P- selectin-mediated event. J Immunol 1994; 152(7):3570-3577.
22. Miki I, Kusano A, Ohta S et al. Histamine enhanced the TNF-[alpha]-induced expression of E-selectin and ICAM-1 on vascular endothelial cells. Cell Immunol 1996; 171(2):285-288.
23. Molet S, Gosset P, Lassalle P et al. Inhibitory activity of loratadine and descarboxyethoxyloratadine on histamine-induced activation of endothelial cells. Clin Exp Allergy 1997; 27:1167-1174.
24. Delneste Y, Lassalle P, Jeannin P et al. Histamine induces IL-6 production by human endothelial cells. Clin Exp Immunol 1994; 98(2):344-349.
25. Takizawa H, Ohtoshi T, Kikutani T et al. Histamine activates bronchial epithelial cells to release inflammatory cytokines in vitro. Int Arch Allergy Immunol 1995; 108(3):260-267.
26. Arima M, Plitt J, Stellato C et al. Expression of interleukin-16 by human epithelial cells inhibition by dexamethasone. Am J Respir Cell Mol Biol 1999; 21(6):684-692.
27. Gantner F, Sakai K, Tusche MW et al. Histamine H_4 and H_2 receptors control histamine-induced interleukin-16 release from human CD8+ T-cells. J Pharmacol Exp Ther 2002; 303(1):300-307.
28. Panettieri RA, Yadvish PA, Kelly AM et al. Histamine stimulates proliferation of airway smooth muscle and induces c-fos expression. American J Physiol 1990; 259(6, Pt 1):L365-L371.
29. Shelhamer JH, Marom Z, Kaliner M. Immunologic and neuropharmacologic stimulation of mucous glycoprotein release from human airways in vitro. J Clin Invest 1980; 66(6):1400-1408.
30. Hirata N, Takeuchi K, Ukai K et al. Expression and localization of histamine H_2 receptor messenger RNA in human nasal mucosa. J Allergy Clin Immunol 1999; 103(5):944-949.
31. Bradding P, Walls AF, Holgate ST. The role of the mast cell in the pathophysiology of asthma. The J Allergy Clin Immunol 2006; 117(6):1277-1284.
32. Hofstra CL, Desai PJ, Thurmond RL et al. Histamine H_4 receptor mediates chemotaxis and calcium mobilization of mast cells. J Pharmacol Exp Ther 2003; 305(3):1212-1221.
33. Godot V, Arock M, Garcia G et al. H_4 histamine receptor mediates optimal migration of mast cell precursors to CXCL12. J Allergy Clin Immunol 2007; 120(4):827-834.
34. Thurmond RL, Desai PJ, Dunford PJ et al. A potent and selective histamine H_4 receptor antagonist with anti-inflammatory properties. J Pharmacol Exp Ther 2004; 309(1):404-413.
35. Flint KC, Leung KB, Hudspith BN et al. Bronchoalveolar mast cells in extrinsic asthma: a mechanism for the initiation of antigen specific bronchoconstriction. Br Med J 1985; 291:923-926.
36. MacGlashan D. Histamine: A mediator of inflammation. J Allergy Clin Immunol 2003; 112(4):S53-S59.
37. Lichtenstein LM, Gillespie E. Effects of the H_1 and H_2 antihistamines on allergic histamine release and its inhibition by histamine. J Pharmacol Exp Ther 1975; 192(2):441-450.

38. Lippert U, Moller A, Welker P et al. Inhibition of cytokine secretion from human leukemic mast cells and basophils by H_1- and H_2-receptor antagonists. Exp Dermatol 2000; 9(2):118-124.
39. Haldar P, Brightling CE, Hargadon B et al. Mepolizumab and exacerbations of refractory eosinophilic asthma. N Engl J Med 2009; 360(10):973-984.
40. Nair P, Pizzichini MMM, Kjarsgaard M et al. Mepolizumab for prednisone-dependent asthma with Sputum eosinophilia. N Engl J Med 2009; 360(10):985-993.
41. Buckland KF, Williams TJ, Conroy DM. Histamine induces cytoskeletal changes in human eosinophils via the H_4 receptor. Br J Pharmacol 2003; 140(6):1117-1127.
42. Ling P, Ngo K, Nguyen S et al. Histamine H_4 receptor mediates eosinophil chemotaxis with cell shape change and adhesion molecule upregulation. Br J Pharmacol. 2004; 142(1):161-171.
43. Maneechotesuwan K, Essilfie-Quaye S, Kharitonov SA et al. Loss of control of asthma following inhaled corticosteroid withdrawal is associated with increased sputum interleukin-8 and neutrophils. Chest 2007; 132(1):98-105.
44. Wenzel Sally E, Szefler Stanley J, Leung Donald YM et al. Bronchoscopic evaluation of severe asthma. Persistent inflammation associated with high dose glucocorticoids. Am J Respir Crit Care Med 1997; 156(3):737-743.
45. Shannon J, Ernst P, Yamauchi Y et al. Differences in airway cytokine profile in severe asthma compared to moderate asthma. Chest 2008; 133(2):420-426.
46. Lamblin C, Gosset P, Tillie-Leblond I et al. Bronchial neutrophilia in patients with noninfectious status asthmaticus. Am J Respir Crit Care Med 1998; 157(2):394-402.
47. Xu X, Zhang D, Zhang H et al. Neutrophil histamine contributes to inflammation in mycoplasma pneumonia. J Exp Med 2006; 203(13):2907-2917.
48. Flamand N, Plante H, Picard S et al. Histamine-induced inhibition of leukotriene biosynthesis in human neutrophils: involvement of the H_2 receptor and cAMP. Br J Pharmacol 2004; 141(4):552-561.
49. Takeshita K, Sakai K, Bacon KB et al. Critical role of histamine H_4 receptor in leukotriene B4 production and mast cell-dependent neutrophil recruitment induced by zymosan in vivo. J Pharmacol Exp Ther 2003; 307(3):1072-1078.
50. Takeshita K, Bacon KB, Gantner F. Critical role of L-selectin and histamine H4 receptor in zymosan-induced neutrophil recruitment from the bone marrow: Comparison with carrageenan. J Pharmacol Exp Ther 2004; 310(1):272-280.
51. Spiteri MA, Knight RA, Jeremy JY et al. Alveolar macrophage-induced suppression of peripheral blood mononuclear cell responsiveness is reversed by in vitro allergen exposure in bronchial asthma. Eur Respir J 1994; 7(8):1431-1438.
52. St-Laurent J, Turmel Vr, Boulet L-P et al. Alveolar macrophage subpopulations in bronchoalveolar lavage and induced sputum of asthmatic and control subjects. J Asthma 2009; 46(1):1-8.
53. Elenkov IJ, Webster E, Papanicolaou DA et al. Histamine potently suppresses human IL-12 and stimulates IL-10 production via H_2 receptors. J Immunol 1998; 161(5):2586-2593.
54. Van der Pouw Kraan TCTM, Snijders A, Boeije LCM et al. Histamine inhibits the production of interleukin-12 through interaction with H_2 receptors. J Clin Invest 1998; 102(10):1866-1873.
55. Dijkstra D, Leurs R, Chazot P et al. Histamine downregulates monocyte CCL2 production through the histamine H_4 receptor. J Allergy Clin Immunol 2007; 120(2):300-307.
56. Triggiani M, Gentile M, Secondo A et al. Histamine induces exocytosis and IL-6 production from human lung macrophages through interaction with H_1 receptors. J Immunol 2001; 166(6):4083-4091.
57. Hammad H, Lambrecht BN. Recent progress in the biology of airway dendritic cells and implications for understanding the regulation of asthmatic inflammation. J Allergy Clin Immunol 2006; 118(2):331-336.
58. Szeberenyi JB, Pallinger E, Zsinko M et al. Inhibition of effects of endogenously synthesized histamine disturbs in vitro human dendritic cell differentiation. Immunol Lett 2001; 76(3):175-182.
59. Mazzoni A, Young HA, Spitzer JH et al. Histamine regulates cytokine production in maturing dendritic cells, resulting in altered T-cell polarization. J Clin Invest 2001; 108(12):1865-1873.
60. Gutzmer R, Langer K, Lisewski M et al. Expression and function of histamine receptors 1 and 2 on human monocyte-derived dendritic cells. J Allergy Clin Immunol 2002; 109(3):524-531.
61. Gutzmer R, Diestel C, Mommert S et al. Histamine H_4 receptor stimulation suppresses IL-12p70 production and mediates chemotaxis in human monocyte-derived dendritic cells. J Immunol 2005; 174(9):5224-5232.
62. Dunford PJ, O'Donnell N, Riley JP et al. The histamine H_4 receptor mediates allergic airway inflammation by regulating the activation of CD4+ T-cells. J Immunol 2006; 176(11):7062-7070.
63. Jutel M, Klunker S, Akdis M et al. Histamine upregulates Th1 and downregulates Th2 responses due to different patterns of surface histamine 1 and 2 receptor expression. Int Arch Allergy Immunol 2001; 124(1-3):190-192.

64. Gutzmer R, Mommert S, Gschwandtner M et al. The histamine H_4 receptor is functionally expressed on TH2 cells. J Allergy Clin Immunol 2009; 123(3):619-625.
65. Jutel M, Watanabe T, Klunker S et al. Histamine regulates T-cell and antibody responses by differential expression of H_1 and H_2 receptors. Nature 2001; 413(6854):420-425.
66. Banu Y, Watanabe T. Augmentation of antigen receptor-mediated responses by histamine H_1 receptor signaling. J Exp Med 1999; 189(4):673-682.
67. Kunzmann S, Mantel PY, Wohlfahrt JG et al. Histamine enhances TGF-b1-mediated suppression of Th2 responses. FASEB 2003; 17(9):1089-1095.
68. Sonobe Y, Nakane H, Watanabe T et al. Regulation of con A-dependent cytokine production from CD4+ and CD8+ T-lymphocytes by autosecretion of histamine. Inflamm Res 2004; 53(3):87-92.
69. Bryce PJ, Mathias CB, Harrison KL et al. The H_1 histamine receptor regulates allergic lung responses. J Clin Invest 2006; 116(6):1624-1632.
70. Morgan RK, McAllister B, Cross L et al. Histamine 4 receptor activation induces recruitment of FoxP3+ T-cells and inhibits allergic asthma in a murine model. J Immunol 2007; 178(12):8081-8089.
71. Leite-De-Moraes MC, Moreau G, Arnould A et al. IL-4-producing NK T-cells are biased towards IFN-and ggr; production by IL-12. Influence of the microenvironment on the functional capacities of NK T-cells. European J Immunol 1998; 28(5):1507-1515.
72. Agea E, Russano A, Bistoni O et al. Human CD1-restricted T-cell recognition of lipids from pollens. J Exp Med 2005; 202(2):295-308.
73. Leite-de-Moraes MC, Diem S, Michel M-L et al. Cutting edge: histamine receptor H_4 activation positively regulates in vivo IL-4 and IFN-gamma production by invariant NKT cells. J Immunol 2009; 182(3):1233-1236.
74. Blumchen K, Gerhold K, Thorade I et al. Oral administration of desloratadine prior to sensitization prevents allergen-induced airway inflammation and hyper-reactivity in mice. Clin Exp Allergy 2004; 34(7):1124-1130.
75. Gelfand EW, Cui Z-H, Takeda K et al. Fexofenadine modulates T-cell function, preventing allergen-induced airway inflammation and hyperresponsiveness. J Allergy Clin Immunol 2002/7 2002; 110(1):85-95.
76. Siebenhaar F, Degener F, Zuberbier T et al. High-dose desloratadine decreases wheal volume and improves cold provocation thresholds compared with standard-dose treatment in patients with acquired cold urticaria: a randomized, placebo-controlled, crossover study. J Allergy Clin Immunol 2009; 123(3):672-679.
77. Miyamoto K, Iwase M, Nyui M et al. Histamine type 1 receptor deficiency reduces airway inflammation in a murine asthma model. Int Arch Allergy Immunol 2006; 140(3):215-222.
78. Deml K-F, Beermann S, Neumann D et al. Interactions of histamine H_1-receptor agonists and antagonists with the human histamine H_4-receptor. Mol Pharmacol 2009; 76(5):1019-1030.
79. ETAC. Allergic factors associated with the development of asthma and the influence of cetirizine in a double-blind, randomised, placebo-controlled trial: first results of ETAC. Early development of the atopic child. Pediatr Allergy Immunol 1998; 9:116-124.
80. De Bie JJ, Henricks PAJ, Cruikshank WW et al. Modulation of airway hyperresponsiveness and eosinophilia by selective histamine and 5-HT receptor antagonists in a mouse model of allergic asthma. Br J Pharmacol 1998; 124(5):857-864.
81. Leopold JD, Hartley JP, Smith AP. Effects of oral H_1 and H_2 receptor antagonists in asthma. Br J Clin Pharmacol 1979; 8(3):249-251.
82. Nogrady SG, Hahn AG. H_2-receptor blockade and exercise-induced asthma. Br J Clin Pharmacol 1984; 18(5):795-797.
83. Popa VT. Bronchodilating activity of an H_1 blocker, chlorpheniramine. J Allergy Clin Immunol 1977; 59(1):54-63.
84. Popa VT. Effect of an H1 blocker, chlorpheniramine, on inhalation tests with histamine and allergen in allergic asthma. Chest 1980; 78(3):442-451.
85. Hodges IG, Milner AD, Stokes GM. Bronchodilator effect of two inhaled H1-receptor antagonists, clemastine and chlorpheniramine, in wheezy school-children. Br J Dis Chest 1983; 77(3):270-275.
86. Dorward AJ, Patel KR. A comparison of ketotifen with clemastine, ipratropium bromide and sodium cromoglycate in exercise-induced asthma. Clin Allergy 1982; 12(4):355-361.
87. Groggins RC, Milner AD, Stokes GM. Bronchodilator effects of clemastine, ipratropium, bromide and salbutamol in preschool children with asthma. Arch Dis Child 1981; 56(5):342-344.
88. Henry RL, Hodges IG, Milner AD et al. Bronchodilator effects of the H_1 receptor antagonist—clemastine. Arch Dis Child 1983; 58(4):304-305.
89. Van Ganse E, Kaufman L, Derde MP et al. Effects of antihistamines in adult asthma: a meta-analysis of clinical trials. Eur J Respir Dis 1997; 10(10):2216-2224.

90. Chan TB, Shelton DM, Eiser NM. Effect of an oral H_1-receptor antagonist, terfenadine, on antigen-induced asthma. Br J Dis Chest 1986; 80(4):375-384.
91. Eiser NM. The effect of a beta 2-adrenergic agonist and a histamine H_1-receptor antagonist on the late asthmatic response to inhaled antigen. Respir Med 1991; 85(5):393-399.
92. Hamid M, Rafferty P, Holgate S. The inhibitory effect of terfenadine and flurbiprofen on early and late-phase bronchoconstriction following allergen challenge in atopic asthma. Clin Exp Allergy 1990; 20(3):261-267.
93. Taytard A, Beaumont D, Pujet JC et al. Treatment of bronchial asthma with terfenadine; a randomized controlled trial. Br J Clin Pharmacol 1987; 24(6):743-746.
94. Rafferty P, Jackson L, Smith R et al. Terfenadine, a potent histamine H_1-receptor antagonist in the treatment of grass pollen sensitive asthma. Br J Clin Pharmacol 1990; 30(2):229-235.
95. Wood-Baker R, Smith R, Holgate ST. A double-blind, placebo controlled study of the effect of the specific histamine H_1-receptor antagonist, terfenadine, in chronic severe asthma. Br J Clin Pharmacol 1995; 39(6):671-675.
96. Kaliner M, Gottlieb G, Lieberman A et al. Safety of fexofenadine HCI in asthmatic patients requiring inhaled corticosteroids in accordance with NHLBI guidlines. J Allergy Clin Immunol 2000; 105(1, Part 2):S388.
97. Spector S, Lee N, McNutt B et al. Effect of terfenadine in asthmatic patients. Ann Allergy 1992; 69(3):212-216.
98. Bruttmann G, Pedrali P, Arendt C et al. Protective effect of cetirizine in patients suffering from pollen asthma. Ann Allergy 1990; 64(2):224-228.
99. Corren J, Harris AG, Aaronson D et al. Efficacy and safety of loratadine plus pseudoephedrine in patients with seasonal allergic rhinitis and mild asthma. J Allergy Clin Immunol 1997; 100(6):781-788.
100. Reicin A, White R, Weinstein SF et al. Montelukast, a leukotriene receptor antagonist, in combination with loratadine, a histamine receptor antagonist, in the treatment of chronic asthma. Arch Intern Med 2000; 160(16):2481-2488.
101. Ekstrom T, Osterman K, Zetterstrom O. Lack of effect of loratadine on moderate to severe asthma. Ann Allergy Asthma Immunol 1995; 75(3):287-289.
102. Gelfand EW, Cui Z-H, Takeda K et al. Effects of fexofenadine on T-cell function in a murine model of allergen-induced airway inflammation and hyperresponsiveness. J Allergy Clin Immunol 2003/10 2003; 112(4, Supplement 1):S89-S95.
103. Bousquet J, Emonot A, Germouty J et al. Double-blind multicenter study of cetirizine in grass-pollen-induced asthma. Ann Allergy 1990; 65(6):504-508.
104. Wood-Baker R, Smith R, Holgate ST. A double-blind, placebo controlled study of the effect of the specific histamine H_1-receptor antagonist, terfenadine, in chronic severe asthma. Br J Clin Pharmacol 1995; 39(6):671-675.

Chapter 7

Antihistamines in the Treatment of Urticaria

Torsten Zuberbier* and Marcus Maurer*

Abstract

Most urticaria subtypes have a profound impact on patients' quality of life and performance. Effective treatment is thus required in all cases where the avoidance of eliciting factors and the elimination of underlying causes is not feasible. In nearly all subtypes histamine released by mast cells plays a predominant role. For symptomatic relief second generation nonsedating histamine H_1 receptor (H_1R)-antihistamines are, therefore, the first choice. However, double-blind controlled studies have shown that dosages required to achieve complete protection from urticaria symptoms may exceed those recommended for other diseases, e.g., allergic rhinitis. Therefore, the current guidelines suggest increasing the dosage up to fourfold, whereas alternative treatments should be reserved as add-on therapy for unresponsive patients.

Introduction

Urticaria as a disease was described as far back as Hippocrates, but the many different types and subtypes of urticaria have only been recognized and characterized during the past century. With an increasing understanding of the molecular mechanisms involved in urticaria pathogenesis, there is also growing evidence for the heterogeneity of the disease. This new knowledge calls for an individualized approach, which is especially important since the impairment of the quality of life in urticaria is often high. For example, Poon et al[1] report that chronic spontaneous urticaria patients suffer an impairment of health that is similar to that seen in patients with psoriasis or acne in a study of 170 consecutive patients attending a specialist urticaria clinic.

According to the new EAACI (European Academy of Allergology and Clinical Immunology) Dermatology Section, the EU-funded network of excellence, GA^{2}LEN (Global Allergy and Asthma European Network), the EDF (European Dermatology Forum) and WAO (World Allergy Organization) guidelines the different urticaria subtypes can be grouped into spontaneous urticaria, which includes acute urticaria and chronic urticaria, the physical urticarias and other urticarias including contact urticaria, e.g., wheals and angioedema formation that is directly linked to the degranulation of mast cells either in superficial layers (wheals) or in deeper layers of the skin (angioedema). Because of the many different subtypes of urticaria a clear definition and classification of the disease is a prerequisite for a suitable treatment. The clinical appearance is defined as follows:

- Urticaria is characterized by the rapid appearance of wheals and/or angioedema.
- A wheal consists of three typical features:
  1. a central swelling of variable size,
  2. an associated itching or sometimes burning, and
  3. a fleeting duration of usually 1-24 hours.

*Correspondence—Allergie-Centrum-Charité, Charité-Universitätsmedizin Berlin, Chariteplatz 1, D-10117 Berlin. Email: torsten.zuberbier@charite.de, marcus.maurer@charite.de

Histamine in Inflammation, edited by Robin L. Thurmond.

Table 1. Classification of urticaria[30]

Type	Subtype	Definition
Spontaneous urticaria	Acute spontaneous urticaria	Spontaneous wheals and/or angioedema <6 weeks
	Chronic spontaneous urticaria	Spontaneous wheals and/or angioedema >6 weeks
Physical urticaria	Cold contact urticaria	Eliciting factor: cold objects/air/fluids/wind
	Delayed pressure urticaria	Eliciting factor: vertical pressure (wheals arising with a 3-12 h latency)
	Heat contact urticaria	Eliciting factor: localized heat
	Solar urticaria	Eliciting factor: UV and/or visible light
	Symptomatic dermographism/ Urticaria factitia	Eliciting factor: mechanical shearing forces (wheals arising after 1-5 min)
	Vibratory urticaria/angioedema	Eliciting factor: vibratory forces, e.g., pneumatic hammer
Other urticaria disorders	Aquagenic urticaria	Eliciting factor: water
	Cholinergic urticaria	Elicitation by increase of body core temperature due to physical exercises, spicy food
	Contact urticaria	Elicitation by contact with urticariogenic substance
	Exercise induced anaphylaxis/ urticaria	Eliciting factor: physical exercise

Angioedema is defined by:

- Sudden, pronounced swelling of the lower dermis and subcutis.
- Possible pain.
- Resolution is slower than for wheals (up to 72 hours).

Nearly all symptoms of urticaria are mediated by histamine as the major proinflammatory mast cell mediator binding to H_1R on endothelial cells and sensory nerves. However, the clinical manifestations of different urticaria subtypes vary considerably. Also, it is important to note that in one patient two or more different subtypes of urticaria can coexist. Table 1 presents a classification for clinical use.

Physical urticarias, although of a chronic nature, are grouped separately since they depend on the presence of their eliciting physical factors, whereas in acute and chronic spontaneous urticaria wheals arise spontaneously without external physical stimuli.

Another important factor in classifying urticaria is disease activity. Where physical triggers are implicated, e.g., temperature, an exact measurement of the intensity of the eliciting factor can be made. However for spontaneous urticaria, assessing disease activity is more complex. The new guidelines propose a unified scoring system that will facilitate comparison of study results from different centers. This simple scoring system is based on the assessment of key urticaria symptoms (wheals and pruritus). The self-evaluation of the last 24 hours each day by the patient has proven to be very robust and helpful since disease activity often varies during the day.

Acute Urticaria

For acute urticaria life-time prevalence ranges from 12-15%[2,3] or even, in one study, 23.5%.[4] In a prospective study, in a rural area of Brandenburg in Germany, a one-year incidence of 0.154% was found, which equals a life-time prevalence of 12.32% based on a life expectancy of 80 years.[5,6] However, mild symptoms may not have been reported and the true life-time prevalence must be estimated to be rather 15-20%.

Regarding the etiology of the disease, the above-mentioned prospective study in acute urticaria showed that although 63% of the patients suspected food to be the cause, in only 1 of 109 patients was food shown to be the causing agent upon thorough investigation. This highlights that patient history, especially in acute urticaria, may be misleading.[6] Drugs can elicit acute urticaria both as allergens (e.g., penicillin) and as pseudoallergens (e.g., NSAID). The most frequent reason for acute urticaria, however, appears to be viral infections of the upper respiratory tract.

Chronic Spontaneous Urticaria

Due to the lack of cross-sectional studies, there is no reliable data regarding the prevalence of chronic spontaneous urticaria although it is estimated to be 1%. As for acute spontaneous urticaria, Type I—allergic reactions are only rarely responsible for the development of chronic spontaneous urticaria.[7,8] In different subsets of patients with chronic spontaneous urticaria, the role of pseudoallergic reactions against food and food additives have been repeatedly discussed in the past as well as an infectious or autoreactive etiology.

Our own results[8] show that in those patients who improve on a diet low in pseudoallergens, 30% show a decrease of symptoms only after 10-14 days on the diet. The study included unselected chronic spontaneous urticaria patients with daily or almost daily symptoms who had not received a diagnostic work-up before. Approximately 50% of the responders did not express a total clearance of symptoms, pointing at other possible cofactors involved in the pathogenesis. These results were confirmed by Pigatto and Valsecchi[9] who investigated a group of 202 patients with chronic spontaneous urticaria using the same diet. In this study 126 patients improved on diet, whereas 35 patients did not show any benefit from the diet and 41 patients dropped out. In both studies reactions to food additives were only seen in a minority of patients (19% and 37%, respectively). Meanwhile the relevance of naturally occurring pseudoallergens, especially aromatic compounds found in vegetables and wine, have been confirmed.[10] Various studies have investigated the occurrence of anti-FcεRIα autoantibodies, which have been described to be of pathophysiological relevance in some patients with urticaria.[11-13] In our own study these autoantibodies were found in the same frequency as described earlier, but they could be found in both patients with idiopathic chronic spontaneous urticaria (7 of 22) and in patients with pseudoallergy against food whose symptoms cleared with elimination diet (6 of 17).[14] In addition, it has been shown that these autoantibodies crosslink the IgE-receptor only if it is not occupied by IgE, which is rarely the case under physiological conditions.[15,16] Two possible explanations for these findings are conceivable. First, the FcεRIα autoantibodies are not of pathophysiological relevance in all patients with urticaria or second, a synergism between the autoantibodies and other eliciting stimuli, e.g., food, is necessary for the appearance of clinical symptoms in some patients. Future research in this field is necessary to answer this question.

Apart from anti-IgE receptor autoantibodies, thyroid autoantibodies are also associated with chronic spontaneous urticaria,[17] although the pathomechanism is unclear. Infections such as hepatitis A and B, bacterial infections, e.g., of the nasopharynx or helicobacter pylori of the gastrointestinal tract,[18-21] can also trigger chronic spontaneous urticaria and should be treated appropriately.

Parasites, a rare cause of urticaria in North European countries but more frequent in other regions, should be eliminated. In the past, intestinal candidosis has been regarded as a highly important eliciting factor for chronic spontaneous urticaria,[3] but recent findings fail to support a significant causative role.[6] Nevertheless, it is recommended that massive candidosis should be treated. In general, the frequency and relevance of infectious diseases as a cause for chronic urticaria varies between different patient groups and in different regions. For example, hepatitis virus infections are a more frequent cause for chronic spontaneous urticaria in southern Europe, but a rare cause in northern Europe. Apart from infections, also non-infectious chronic inflammatory processes such as gastritis, reflux esophagitis, inflammation of the bile duct or bile gland,[6,22] or rarely autoimmune disorders, e.g., SLE, have been identified to cause urticaria in some patients.

Management of Urticaria Follows Basic Principles

The management of urticaria should follow the newly developed EAACI/GA^{2}LEN/EDF/WAO-Guideline.[23] This guideline is the result of a consensus reached during a panel discussion at the 3nd International Consensus Meeting on Urticaria, *Urticaria 2008*, a joint initiative of EAACI Dermatology Section, the EU-funded network of excellence, GA^{2}LEN, the EDF and WAO.

Although urticaria is elicited by a great diversity of factors and clinically presents in a highly variable way its treatment mainly follows the same basic principles. These are:

- Avoidance or elimination of the eliciting stimulus or underlying cause.
- Symptomatic treatment mainly targeting either reducing mast cell mediator release or reducing the effect of these mediators on the target organs.

In all cases symptomatic relief should be offered while searching for causes. Avoidance, elimination or treatment of the eliciting stimulus or cause is most desirable since it is curative, but unfortunately it is not applicable in the majority of patients as the exact eliciting stimulus is frequently unknown.

Antihistamines Play the Major Role in Urticaria Treatment

Nearly all symptoms of urticaria are mediated primarily by the actions of histamine on H_1R located on endothelial cells (the wheal) and on sensory nerves (neurogenic flare and pruritus). Thus, H_1R antagonists are of eminent importance in the treatment of urticaria. The availability of this class of therapeutics since the 1950s has made urticaria a disease that can be treated effectively with a very low adverse effect profile. The older 1st generation antihistamines have pronounced CNS and anticholinergic effects which last longer than 12 hours, whereas the antipruritic effect lasts only for 4-6 hours. Consequently, many drug interactions have been described for these sedating antihistamines, particularly with drugs affecting the central nervous system, like analgesics, hypnotics, sedatives and mood elevating drugs as well as alcohol. Also, MAO inhibitors can prolong and intensify the anticholinergic effects of these drugs. In addition, 1st generation antihistamines can interfere with REM sleeping phases and impact learning and presentation.

The development of 2nd generation antihistamines led to drugs that are nonsedating or minimally-sedating and are free of anticholinergic effects. However, two of the earlier 2nd generation drugs, astemizole and terfenadine, which were essentially pro-drugs requiring hepatic metabolism to become fully active, had cardiotoxic effects if this metabolism was blocked by concomitant administration of ketokonazole or erythromycin. These two drugs are no longer available in most countries. Further progress with regard to drug safety was achieved by the development of the new generation antihistamines fexofenadine and desloratadine, which are cytochrome P450 independent metabolites of earlier antihistamines. Levocetirizine is the active enantiomer of cetirizine, thus, where cetirizine is indicated as effective treatment, levocetirizine could also be considered.

Thus, considering their good safety profile, 2nd generation antihistamines must be considered as first line symptomatic treatment for urticaria. However, as of now well-designed randomized controlled clinical trials comparing the efficacy and safety of different nonsedating H_1R-antihistamines in chronic urticaria are largely missing. However, while single dose is sufficient in up to 50% of patients with chronic spontaneous urticaria, many patients need additional therapy. There are some studies showing the benefit of a higher dosage of antihistamines in individual patients.[24,25] This has been verified in studies using up to fourfold higher than recommended doses of desloratadine.[26] Interestingly, however, Asero[27] reported that increasing the dose of cetirizine for chronic spontaneous urticaria threefold did not produce further efficacy in severely affected patients. Most likely the increase of dosage does not only block histamine mediated effects, but also reduces mast cell activation and has an impact on various cytokine and endothelial adhesion molecules. The highest reported accidental overdosage of antihistamine (fifty-fold the prescribed dosage of cetirizine in an 18-month-old boy) induced no adverse effects.[28]

Further Therapeutic Possibilities Are Limited

Alternative treatments are needed for patients unresponsive to higher dosages of antihistamines. Although the mechanism remains unclear, it can be speculated that individual profiles of other mediators and histamine receptors other than the H_1R are involved. Since the side-effects of many of these substances are considerable, it may be wise to try to use them as add-on therapy only in patients unresponsive to antihistamines. Since the severity of urticaria may fluctuate and since spontaneous remission may occur at any time, it is recommended that the necessity for continued or alternative drug treatment should be re-evaluated every 3-6 months.

Modern Antihistamines Should also Be Used for Children

Many clinicians use 1st generation H_1R-antihistamines as their first choice in the treatment of children with allergies assuming that the safety profile of these drugs is better known than that of the 2nd generation nonsedating H_1R-antihistamines due to a longer life on the market. Also the use of 2nd generation H_1R-antihistamines is restricted for children less than 6 months of age, while the recommendation for the 1st generation H_1R-antihistamines is sometimes less clear as these drugs have been licensed in a period when the code of Good Clinical Practice for pharmaceutical industry was less stringent. As a consequence many doctors choose a 1st generation antihistamine, which as pointed out above have a lower safety profile compared with a 2nd generation H_1R-antihistamine. This practice must be discouraged. Thus, in children the same first line treatment and updosing (weight adjusted) is recommended as in adults.

The same considerations in principle apply to pregnant and lactating women. So far no reports of birth defects in women having used 2nd generation antihistamines during pregnancy have been published. Although only small sample size studies are available[29] it must be assumed that due to the wide use of 2nd generation antihistamines in allergic rhinitis and urticaria many women have used these drugs especially in the beginning of pregnancy, at least before the pregnancy was confirmed and most likely later since a number of these drugs do not require a prescription.

Conclusion

Modern H^1R-antihistamines are the mainstay in the treatment of urticaria. However, until now it is not clear if the mode of action in increased dosages is related only to the inactivation of the H^1R-receptor response. Due to the many other anti-inflammatory properties of modern H^1R-antihistamines which are not found in antihistamines of the first generation it is more likely that the positive effects of up-dosing are based on extended modes of action.

Further research is needed to better understand this as well as the role of other histamine receptors than H^1R-receptors in the pathogenesis of urticaria.

References

1. Poon E, Seed PT, Greaves MW et al. The extent and nature of disability in different urticarial conditions. Br J Dermatol 1999; 140:667-671.
2. Sheldon JM, Mathews KP, Lovell RG. The vexing urticaria problem. Present concepts of etiology and management. J Allergy 1954; 25:525-560.
3. Champion RH, Roberts SOB, Carpenter RG et al. Urticaria and angioedema: a review of 554 patients. Br J Derm 1969; 81:588-597.
4. Swinny B. The atopic factor in urticaria. South Med J 1941; 34:855-858.
5. Iffländer J. Akute Urtikaria—Ursachen, Verlauf und Therapie. Berlin, Humboldt University Diss. (Med Doct) 1999.
6. Zuberbier T, Iffländer J, Semmler C et al. Acute urticaria—clinical aspects and therapeutical responsiveness. Acta Derm Venereol (Stockh) 1996; 76:295-297.
7. Juhlin L. Recurrent urticaria: clinical investigation of 330 patients. Br J Dermatol 1981; 104:369-381.
8. Zuberbier T, Chantraine-Hess S, Hartmann K et al. Pseudoallergen-free diet in the treatment of chronic urticaria—a prospective study. Acta Derm Venereol (Stockh) 1995; 75:484-487.
9. Pigatto PD, Valsecchi RH. Chronic urticaria: a mystery. Allergy 2000; 55:306-30.
10. Zuberbier T, Pfrommer C, Specht K et al. Aromatic components of food as novel eliciting factors of pseudoallergic reactions in chronic urticaria. J Allergy Clin Immunol 2002; 109:348-349.

11. Hide M, Francis DM, Grattan CEH et al. Autoantibodies against the high-affinity IgE receptor as a cause of histamine release in chronic urticaria. N Engl J Med 1993; 328:1599-1604.
12. Fiebiger E, Hammerschmid F, Stingl G et al. Anti-FcεRI-α autoantibodies in autoimmune-mediated disorders. Identification of a structure-function relationship. J Clin Invest 1998; 101:243-251.
13. Kikuchi Y, Kaplan AP. Mechanisms of autoimmune activation of basophils in chronic urticaria. J Allergy Clin Immunol 2001; 107:1056-1062.
14. Zuberbier T, Fiebiger E, Maurer D et al. Anti-FcεRIα serum autoantibodies in different subtypes of urticaria. Allergy 2000; 55:951-954.
15. Stadler BM, Pachlopnik J, Vogel M et al. Conditional autoantibodies in urticaria patients: a unifying hypothesis. J Investig Dermatol Symp Proc 2001; 6:150-152.
16. Horn MP, Pachlopnik JM, Vogel M et al. Conditional autoimmunity mediated by human natural anti-Fc(epsilon)RIalpha autoantibodies? FASEB J 2001; 15:2268-2274.
17. Leznoff A, Sussman GL. Syndrome of idiopathic chronic urticaria and angioedema with thyroid autoimmunity: a study of 90 patients. J Allergy Clin Immunol 1989; 84:66-71.
18. Wedi B, Kapp A. Helicobacter pylori infection and skin diseases. J Physiol Pharmacol 1999; 50:753-776.
19. Gaig P, Garcia-Ortega P, Enrique E et al. Efficacy of the eradication of Helicobacter pylori infection in patients with chronic urticaria. A placebo-controlled double blind study. Allergol Immunopathol (Madr) 2002; 30:255-258.
20. Schnyder B, Helbing A, Pichler WJ. Chronic idiopathic urticaria: natural course and association with Helicobacter pylori infection. Int Arch Allergy Immunol 1999; 119:60-63.
21. Wedi B, Kapp A. Helicobacter pylori infection in skin diseases: a critical appraisal. Am J Clin Dermatol 2002; 3:273-282.
22. Bruno G, Andreozzi P, Graf U. Exercise-induced urticaria—angioedema syndrome: A role in gastroesophageal reflux. In: Vena GA, Puddu P, eds. Proceedings of the international symposium on urticaria. Bari: Publ Scientif 1998;85-89.
23. Zuberbier T, Asero R, Bindslev-Jensen C, Walter Canonica G, Church MK, Giménez-Arnau AM, Grattan CE, Kapp A, Maurer M, Merk HF, Rogala B, Saini S, Sánchez-Borges M, Schmid-Grendelmeier P, Schünemann H, Staubach P, Vena GA, Wedi B. EAACI/GA2LEN/EDF/WAO guideline: management of urticaria. Allergy 2009; 64: 1427-43.
24. Zuberbier T, Münzberger Ch, Haustein U et al. Double-blind crossover study of high dose cetirizine in cholinergic urticaria. Dermatology 1996; 193:324-327.
25. Kontou-Fili K, Maniakatou G, Demaka P et al. Therapeutic effect of cetirizine 2HCl in delayed pressure urticaria. Health Sci Rev 1989; 3:23-25.
26. Siebenhaar F, Degener F, Zuberbier T et al. High-dose Desloratadine Decreases Wheal Volume and Improves Cold Provocation Thresholds as Compared with Standard Dose Treatment in Patients with Acquired Cold Urticaria: A Randomized, Placebo-Controlled, Crossover Study. J Allergy Clin Immunol 2009 in press.
27. Asero R. Chronic unremitting urticaria: is the use of antihistamines above the licensed dose effective? A preliminary study of cetirizine at licensed and above-licensed doses. Clin Exp Dermatology; 2007; 32:34-38.
28. Ridout SM, Tariq SM. Cetirizine overdose in a young child. J Allergy Clin Immunol 1997; 99:860-861.
29. Weber-Schoendorfer C, Schaefer C. The safety of cetirizine during pregnancy. A prospective observational cohort study. Reprod Toxicol 2008; 26:19-23.
30. Zuberbier T, Asero R, Bindslev-Jensen C, Walter Canonica G, Church MK, Giménez-Arnau A, Grattan CE, Kapp A, Merk HF, Rogala B, Saini S, Sánchez-Borges M, Schmid-Grendelmeier P, Schünemann H, Staubach P, Vena GA, Wedi B, Maurer M. EAACI/GA2LEN/EDF/WAO guideline: definition, classification and diagnosis of urticaria. Allergy 2009; 64: 1417-26.

Chapter 8

Histamine and Antihistamines in Atopic Dermatitis

Jörg Buddenkotte, Marcus Maurer and Martin Steinhoff*

Abstract

Itching (pruritus) is perhaps the most common symptom associated with inflammatory skin diseases and can be a lead symptom of extracutaneous disease (e.g., malignancy, infection, metabolic disorders). In atopic dermatitis itching sensations constitute one of the most prominent and distressing features. The most characteristic response to itching is the scratch reflex: a more or less voluntary, often sub-conscious motor activity, to counteract the itch by slightly painful stimuli. The benefit of a short-termed relieve from itching through this scratch reflex though is counteracted by a simultaneous damage of the epidermal layer of the skin which leads to increased transepidermal water loss and drying, which in turn results in a cycle of more itching and more scratching. A wide range of peripheral itch-inducing stimuli generated within or administered to the skin are able to trigger pruritus, one of them being histamine. Based on early experiments, histamine has been suggested to may play a key role in the pathogenesis of AD. This is reflected by a history for antihistamines in the therapeutic medication of AD patients. Antihistamines are believed to share a common antipruritic effect and therefore are prescribed to the vast majority of AD patient suffering from itch to act alleviating. The level of evidence in support of the benefits of antihistamine treatment, however, is low. To assess the benefit of antihistamines in the treatment of AD in a better way, their mechanisms and specific effects need to be understood more precisely. In particular their precise indication is crucial for successful use. This book chapter will therefore summarize and assess the role of histamine in AD and the efficacy of antihistamines in its treatment based on results of basic research and clinical studies.

Histamine

Imidazolethylamine, better known as histamine, is synthesized from the amino acid L-histidine through oxidative decarboxylation by histidine decarboxylase (HDC) and occurrs in tissues throughout the body.[1] The precise function of histamine is dependent on its tissue-localization: e.g., in the nervous system it operates as a neurotransmitter and in the immune system, gut and skin it serves as a signaling molecule.[1] Since the first description of histamine more than a century ago,[2] many important implications on human health have been assigned to this molecule.[3] This importance may derive in major parts from the central role histamine plays in inflammatory processes. In early cutaneous inflammation some researchers consider histamine to be the "quintessential mediator".[4] In fact, histamine has been described to trigger the characteristic triad of inflammation—redness, wheal, flare—along with pruritus in the skin.[5,6] The sources of cutaneous histamine are reported to be mast cells and keratinocytes, which release the molecule upon stimulation.[7,8]

*Corresponding Author: Martin Steinhoff—Departments of Dermatology and Surgery, University of California San Francisco, 513 Parnassus Ave, Rocm S-1268, 94143 San Francisco, California, USA. Email: martin_steinhoff@web.de

Histamine in Inflammation, edited by Robin L. Thurmond.

Histamine Receptors

Histamine exerts its action through binding to distinct metabotropic histamine receptor subtypes. To date, four histamine receptor subtypes are known and cloned. All four belong to the rhodopsin-like family of G protein-coupled receptors (GPCR). Also several isoforms exist for each receptor subtype which emerge from different transcriptional and posttranscriptional processing. Remarkably, the histamine receptors display a high degree of functional heterogeneity as well as discriminable expression patterns and a distinct utilization of intracellular signaling mechanisms.[1,9,10] Differences in the functionality of histamine receptor splice variants are not reported so far.[3,9,10] In the skin, the expression of two of the histamine receptors, H_1R and H_2R, has been detected[7,11] but most histamine-mediated dermatological effects are attributed to the classical H_1R.[12]

The histamine H_1 receptor (H_1R) is found throughout the whole body. It is expressed by multiple cell types including endothelial cells, epithelial cells, nerve cells, smooth muscles and immune cells.[9] As indicated, H_1R couples to G-proteins. More precisely H_1R mediates its effects through $G\alpha_{q/11}$ and G_i/G_o. Coupling of these G proteins leads to phospholipase C, protein kinase C activation, Ca^{2+} mobilization, inositol phosphate, arachidonic acid and nitric oxide production[1,9] (Fig. 1). Utilizing these effector pathways, histamine is competent to generate a wide range of effects among which roles in inflammation and immune modulation are described best. In the skin, for example, histamine causes vasodilation (reddening), vascular permeability (wheal), flushing and pruritus.[9] In allergic reactions histamine leads to an increase of the cellular immunity and decrease of the humoral immunity.[9] Further effects typical for allergic reactions which are inducible by the histamine H_1R axis are pain, asthmatic bronchoconstriction and

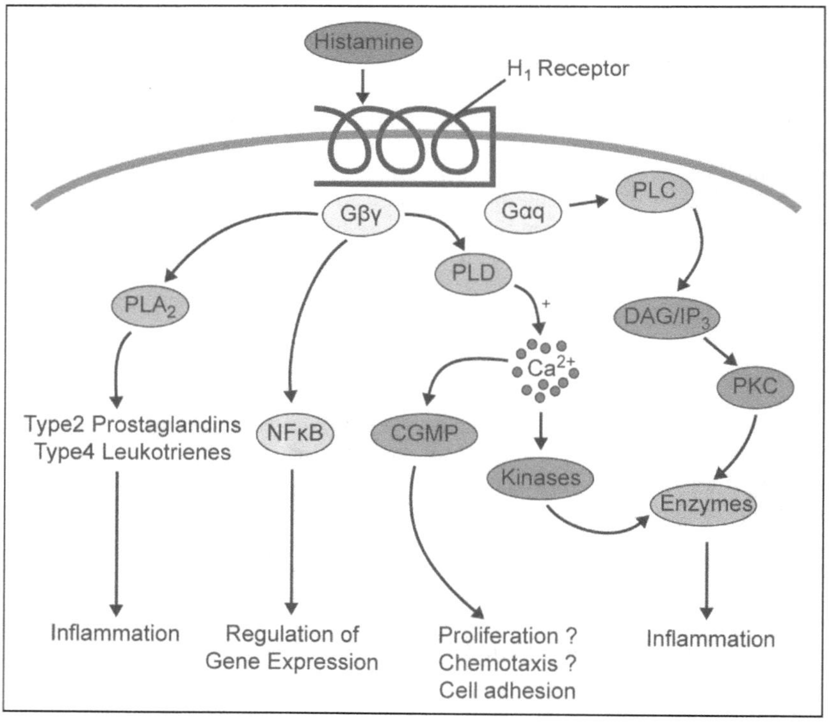

Figure 1. Diagram showing the major G protein-mediated signaling pathways coupled to the histamine H_1 receptor. *Question marks* represent effects of signalling pathways that are not fully revealed to be activated by H_1R, but are in favor of other G protein-coupled receptors or are typical intermediate-accompanying molecules.

coughing. As of yet it is not known which of these H_1R-mediated effects may contribute to the pathology of AD.

Next to H_1R, the histamine H_2 receptor (H_2R) could be a putative candidate receptor that might contribute to AD because its expression pattern closely resembles that of the H_1R including its presence on cutaneous sensory nerve fibers.[7,11] The intracellular effector pathways activated by H_2R are restricted when compared to H_1R effector pathways. Hence, histamine mediates the direct stimulation of membrane adenylate cyclase activity and cAMP-dependent inhibition of cell functions in parts via H_2R, but appears to fail to mediate other major secondary messengers via this receptor.[9] The last cloned histamine receptors are the histamine H_3 (H_3R) and H_4 receptor (H_4R). Due to the expression of the H_3R which does not comprise the epidermal tissue a prominent role for the H_3R in AD or any cutaneous disease is not likely.[13] In contrast, H_4R is coexpressed with H_1R and H_2R by many cells involved in inflammatory processes.[9] Only recently Dilkstra et al provided evidence for the expression of H_4R on inflammatory dendritic epidermal cells (IDEC), which are typically found in lesional skin of patients with AD[14] and revealed an immunomodulatory function for this receptor on this distinct population of dendritic cells. Because of the importance of IDECs in the pathogenesis of AD, the data accumulated by this work for the first time suggests a fundamental role for H_4R in the pathogenesis of ths disease. In mice H_3R as well as the H_4R operate as itch receptors,[15,16] it is likely that both receptors could in part be responsible for the onset of itch also in humans. Antagonists against H_3R and H_4R may in the future become beneficial tools for the treatment of AD symptoms, preferentially for pruritus that does not respond to H_1R targeting compounds.

Histamine in Atopic Dermatitis

Atopic Dermatitis (AD) is a common chronic inflammatory skin disease and is increasing in prevalence throughout the world.[17,18] Patients with AD typically present with xerosis, relapsing eczematous skin lesions and pruritus.[19,20] AD can range from mild disease where only parts of the body are affected to almost total body involvement. There is currently no cure available and therapies are mainly directed at alleviating the symptoms.[19,21] This critical situation mainly results from the poor understanding of the underlying pathogenesis. In the literature one finds indications for a relevant role of histamine in AD, e.g., histamine-free diet reportedly reduces AD symptoms[22,23] and administration of histamine immunoglobuline complexes has been shown to result in clinical improvement.[24] Overall, however, the precise role of histamine in AD remains enigmatic. In the recent scientific debate the role of histamine in AD is mainly seen as that of a molecule with pruritogenic impact.[13,25,26] Earlier studies did find elevated levels of histamine and dermal mast cells in AD patients,[13,25,27,28] topographically associating histamine and itch and underscoring that the itch sensation perceived by AD patients is at least partially due to the action of histamine. However, more recent evidence has not supported this role for histamine in AD.[29] Re-evaluating the potency of histamine to induce itch, it was shown that instead of being pruritogenic, small doses of intracutaneously injected histamine clearly fail to produce itch, but suffice to produce edema and erythema.[30-33] Further studies even report that intracutaneous injection and also iontophoretical application of histamine in AD patients even provokes a reduction of the itch perceived by the proband.[34-37] Additionally, intradermal injection of substance P, a neuropeptide that stimulates histamine release from mast cells, in AD patients produced a reduction in itch perception. This not only emphazises the minor capacity of histamine to induce pruritus in AD,[38] but lets one speculate about the general capacity of histamine as a potent pruritogen. Still there is some speculation that pruritus in AD may be due to centrally rather than peripherally expressed histamine receptors.[13,25] This debate is supported by recent studies where brain activity was measured after histamine-induced itch by neuroimaging methods. The results demonstrated that the brain activity patterns in AD patients and healthy subjects differ and are associated with itch intensity and disease severity.[39,40] Despite of the controversy as to the pruritic nature of histamine, the pro-inflammatory capacity of histamine appears to be assured and may be the major role of histamine in AD.[9]

Antihistamines and Histamine Receptor Antagonism in Atopic Dermatitis

Topical corticosteroids are first-line therapeutic agents and topical calcineurin inhibitors are considered second-line agents for AD patients;[19,21] however, physicians generously prescribe antihistamines to AD patients as an adjunctive therapy. The most commonly prescribed antihistamines in AD are targeted against the classical H_1R.[3,9,13] Antihistamines against the H_2R play a considerably minor role in AD[9] and as of yet, no antihistamines that target the H_3R are reported of to be used in AD patients. Recent results point to a possible role of H_4R in AD but no compound inhibiting this receptor is approved for use in AD patients as of yet.

The H_1R antihistamines applied in the conventional treatment of AD patients are grouped into "first generation" and "second generation" antihistamines. Criteria for grouping are the date of commercial availability, receptor specificity, penetration of the central nervous system and potential for sedation.[3,12,25,41] Earlier generations of H_1 receptor antagonists were associated with a range of unwanted properties. The first generation of drugs were sedative as well as possessing significant anti-cholinergic effects. Second generation drugs did not cross the blood—brain barrier and, therefore, caused less drowsiness. The benefit of first-generation antihistamines in the treatment of AD is described controversially but is often attributed to their anti-pruritic effect that might rely on the sedative property of this group of pharmaceuticals.[25] In fact, clinicians consider the first-generation H_1R antihistamines mainly for the treatment of sleep problems associated with AD.[42] For example, Doxepin, a tricyclic antidepressant, is prescribed to aid the sleep of patients with AD because it has a high H_1R antagonist activity and is sedating.[42]

Interestingly also nonsedative second-generation antihistamines are often approved by the food and drug administration (FDA), especially for the treatment of pruritus associated with AD. A prominent member of this group is fexofenadine. The value of nonsedating second-generation antihistamines in the treatment of AD is not necessarily seen in their anti-pruritic effect but in their anti-inflammatory properties and it appears to be acknowledged that these anti-inflammatory properties are not strictly histamine receptor related.[43,44] For example, cetirizine, a metabolite of hydroxyzine, is successfully deployed in the treatment of AD due to its anti-inflammatory impact: ceterizine inhibits eosinophil chemotaxis, lowers eosinophilic chemokine release and reduces the expression of endothelial adhesion molecules.[45,46] Further examples are loratadine and desloratadine. Both agents modulate inflammatory responses by regulating cytokine release in immune cells such as human mast cells and basophils[47] and by regulating the expression of cellular adhesion molecules of the endothelium.[48] However, it has to be admitted that the prescribed antihistamines, inhibiting H_1R or H_2R, are not very effective to alleviate symptoms of AD, in particular pruritic sensations. Therefore the search for novel antihistaminic drugs targeting e.g., H_3R or H_4R or new treatment regimens is of uttermost importance. A recent study therefore tested the outcome of a combination of antihistamines in a murine model of allergic contact dermatitis which closely resembles an AD phenotype. Typically such mice show a strong hapten-induced scratching behaviour, which in this study was strongly inhibited by a combination of ceterizine and a novel H_4R antagonist named JNJ7777120.[49] These results indicate that combination of antihistamines in fact might become a new option in the treatment of AD. More importantly it strongly suggests a prominent role for H_4R in the pruritus related to AD and along with earlier publications demonstrates that blocking of H_4R relieves such symptoms.[49,50] Another new approach to the treatment of AD patients follows the idea of antagonists that inhibit multiple receptors important to the pathogenesis of the disease alongside of histamine receptors. A promising lead compound fulfilling the criteria is the dual inhibitor YM-344484. It is capable to block both the chemokine receptor 3 (CCR3) induced and H_1R induced influx of Ca^{2+} and upon oral application nearly fully inhibits histamine induced vascular permeability.[51] So far this compound has only been tested in animal models and is far from a use in humans, but due to its strong effect it surely will be an attractive candidate for further investigation with regard to its potential to alleviate symptoms of AD. It has to be noted, that other actions of antihistamines in addition to their anti-inflammatory effects have been discussed to contribute to their ameliorating effect in AD. Another benefit might be the ability of some antihistamines to modulate the status of the skin barrier, which in patients suffering from AD typically is dysfunctional. As a proof of concept, olopatadine hydrochloride, a drug with H_1R antagonistic action, alleviates skin inflammation and pruritic sensations in a murine model of chronic contact dermatitis by accelerating the recovery of the skin barrier function.[52,53]

Clinical Studies of Antihistamines in Atopic Dermatitis

To date, only limited evidence exists that antihistamines are effective in the treatment of AD.[21] Indeed for some antihistamines slight benefits in the amelioration of AD are reported, but such studies are rare and their evidence is limited. In fact despite their clinical success, antihistamines have not been subjected to large, randomized, double-masked, placebo-controlled long-term trials. Only a few quality studies, small randomized controlled or not even randomized small trials exist assessing the efficacy of antihistamine in the treatment in AD.[30,31,54-58] But either these studies pooled ambiguous results or were not able to provide sufficient evidence for a general efficacy of antihistamines in the treatment of AD.[3,12,13,21,25,30] For example in an early small trial, the authors studied the effect of the second-generation antihistamine terfenadine on alleviating AD symptoms and failed to assign a clear beneficial effect for the compound. Still the authors hesitated to disapprove the usage of terfenadine for AD treatment due to inaccuracies of the very own study.[30] In an only recently published trial the authors assessed the efficacy of the second generation antihistamine fexofenadine in AD patients.[59] The main conclusions of this study derived from a population of only twenty patients suffering from a mild form of AD who were randomly assigned in groups of ten receiving the antihistamine and emollient or the antihistamine and a steroid. The sensation of pruritus was evaluated before and after treatment. In fact, the authors were able to monitor an improvement of pruritus in both groups and suggest that fexofenadine is beneficial in the treatment of AD symptoms disregarding the inadequate size of the study groups. Therefore, a strong demand exists to conduct solid, large-scale, placebo-controlled randomized quality studies to finally develop an assessment of the efficacy of antihistamines for AD treatment. Interestingly, these studies provided an insight as to why some antihistamines may be helpful for some patients with AD: the improvements in the clinical condition and patient quality of life may be due primarily to the promotion of restful sleep, rather than to a direct reduction of symptoms.[12,21,25] Therefore, sedation as an unintended side-effect might be the reason why antihistamines are effective in the treatment of AD patients. Nonetheless, not all antihistamines are effective only because they are sedative. Promethazine,[25] chlorpheniramine[31] and clemastine,[30] for examples, are sedative antihistamines but have been found ineffective in clinical trials. Therefore, the effectiveness of some sedative antihistamines has to be independent of a sedative side-effect.

For further reading, a listing summarizing and assessing the available published clinical evidence on the use of antihistamines for AD can be found in the publication of Maurer et al.[60] Although the level of evidence for efficacy of antihistamines in AD is low, clinical experience supports the benefit of non-sedating antihistamines in some AD patients using sometimes higher than standard doses. Furthermore, despite the effectiveness of sedating antihistamines their use should–due to their unfavorable safety profile[61] and especially because of their effects on REM sleep[62]–not be considered as a routine treatment of AD.

Conclusion

Williams[63] suggested that histamine might play a role in the pathogenesis of AD since intramuscular histamine injections resulted in pruritus. Histamine has doubtlessly become one of the most exhaustingly investigated "itchy" agonist, but today it is acknowledged that the chorus of itch-inducing agents contains many other protagonists. Interestingly upon re-evaluation of the role of histamine in patients with AD, it has been shown that histamine accounts for a reduction in itch sensation instead of an enhancement. This observation lets one speculate about the general capacity of histamine as a potent pruritogen in AD. Another observation demonstrates that small doses of histamine are sufficient to produce edema and erythema upon intracutaneous injection and this may be of importance in the pathogenesis of AD. However, certain antihistamines have been shown to be effective for certain subgroups of AD patients. Sedative antihistamines, such as hydroxyzine and cetirizine, appear beneficial for night time use in patients and second generation antihistamines, such as loratadine and desloratadine, appear beneficial for patients with comorbid conditions such as chronic urticaria. Furthermore, novel results demonstrate that H_4R antagonists may have a strong therapeutic utility for treating pruritic diseases in humans which are unaffected by H_1R or H_2R antagonists. There is also hope that a combination of antihistamines might provide a stronger alleviation of pruritus in AD patients. Taken together, although a definite role of histamine in the pathology of AD appears to be difficult to assess, the available data is also not sufficient to eliminate histamine from the list of potential mediators necessary for the onset of AD or AD symptoms, respectively, in particular AD related pruritus. Further research is needed to finally settle the case for histamine and its receptors in AD.

References

1. Haas HL, Sergeeva OA, Selbach O. Histamine in the nervous system. Physiol Rev 2008; 88:1183-241.
2. Dale HH, Laidlaw PP. The physiological action of beta-iminazolylethylamine. J Physiol 1910; 41:318-44.
3. Simons FE. Advances in H1-antihistamines. N Engl J Med 2004; 351:2203-17.
4. Greaves MW, Sabroe RA. Histamine: the quintessential mediator. J Dermatol 1996; 23:735-40.
5. Lewis T. The blood vessels of the human skin and their responses. London: Shaw and Sons, 1927.
6. Lewis T, Grant RT, Marvin HM. Vascular reactions of the skin to injury. Heart 1929; 14:139-60.
7. Hill SJ. Distribution, properties and functional characteristics of three classes of histamine receptor. Pharmacol Rev 1990; 42:45-83.
8. Malaviya R, Morrison AR, Pentland AP. Histamine in human epidermal cells is induced by ultraviolet light injury. J Invest Dermatol 1996; 106:785-9.
9. Akdis CA, Simons FE. Histamine receptors are hot in immunopharmacology. Eur J Pharmacol 2006; 533:69-76.
10. de Esch IJ, Thurmond RL, Jongejan A et al. The histamine H4 receptor as a new therapeutic target for inflammation. Trends Pharmacol Sci 2005; 26:462-9.
11. Stander S, Weisshaar E, Luger TA. Neurophysiological and neurochemical basis of modern pruritus treatment. Exp Dermatol 2008; 17:161-9.
12. O'Donoghue M, Tharp MD. Antihistamines and their role as antipruritics. Dermatol Ther 2005; 18:333-40.
13. Greaves MW. Antihistamines in dermatology. Skin Pharmacol Physiol 2005; 18:220-9.
14. Dijkstra D, Stark H, Chazot PL et al. Human inflammatory dendritic epidermal cells express a functional histamine H4 receptor. J Invest Dermatol 2008; 128:1696-703.
15. Bell JK, McQueen DS, Rees JL. Involvement of histamine H4 and H1 receptors in scratching induced by histamine receptor agonists in Balb C mice. Br J Pharmacol 2004; 142:374-80.
16. Roosterman D, Goerge T, Schneider SW et al. Neuronal control of skin function: the skin as a neuroimmunoendocrine organ. Physiol Rev 2006; 86:1309-79.
17. Williams HC. Epidemiology of atopic dermatitis. Clin Exp Dermatol 2000; 25:522-9.
18. Horii KA, Simon SD, Liu DY et al. Atopic dermatitis in children in the United States, 1997-2004: visit trends, patient and provider characteristics and prescribing patterns. Pediatrics 2007; 120:e527-34.
19. Darsow U, Lubbe J, Taieb A et al. Position paper on diagnosis and treatment of atopic dermatitis. J Eur Acad Dermatol Venereol 2005; 19:286-95.
20. Hanifin JM, Rajka G. Diagnostic features of atopic dermatitis. Acta Derm Venereol 1980; 92:S44-S7.
21. Hanifin JM, Cooper KD, Ho VC et al. Guidelines of care for atopic dermatitis, developed in accordance with the American Academy of Dermatology (AAD)/American Academy of Dermatology Association "Administrative Regulations for Evidence-Based Clinical Practice Guidelines". J Am Acad Dermatol 2004; 50:391-404.
22. Maintz L, Benfadal S, Allam JP et al. Evidence for a reduced histamine degradation capacity in a subgroup of patients with atopic eczema. J Allergy Clin Immunol 2006; 117:1106-12.
23. Worm M, Fiedler EM, Dolle S et al. Exogenous histamine aggravates eczema in a subgroup of patients with atopic dermatitis. Acta Derm Venereol 2009; 89:52-6.
24. Nahm DH, Lee ES, Park HJ et al. Treatment of atopic dermatitis with a combination of allergen-specific immunotherapy and a histamine-immunoglobulin complex. Int Arch Allergy Immunol 2008; 146:235-40.
25. Herman SM, Vender RB. Antihistamines in the treatment of dermatitis. J Cutan Med Surg 2003; 7:467-73.

26. Ikoma A. Analysis of the mechanism for the development of allergic skin inflammation and the application for its treatment: mechanisms and management of itch in atopic dermatitis. J Pharmacol Sci 2009; 110:265-9.
27. Johnson HH, Jr., Deoreo GA, Lascheid WP et al. Skin histamine levels in chronic atopic dermatitis. J Invest Dermatol 1960; 34:237-8.
28. Juhlin L. Localization and content of histamine in normal and diseased skin. Acta Derm Venereol 1967; 47:383-91.
29. Reitamo S, Ansel JC, Luger TA. Itch in atopic dermatitis. J Am Acad Dermatol 2001; 45:S55-6.
30. Klein PA, Clark RA. An evidence-based review of the efficacy of antihistamines in relieving pruritus in atopic dermatitis. Arch Dermatol 1999; 135:1522-5.
31. Munday J, Bloomfield R, Goldman M et al. Chlorpheniramine is no more effective than placebo in relieving the symptoms of childhood atopic dermatitis with a nocturnal itching and scratching component. Dermatology 2002; 205:40-5.
32. Rukwied R, Lischetzki G, McGlone F et al. Mast cell mediators other than histamine induce pruritus in atopic dermatitis patients: a dermal microdialysis study. Br J Dermatol 2000; 142:1114-20.
33. Hanifin JM. The role of antihistamines in atopic dermatitis. J Allergy Clin Immunol 1990; 86:666-9.
34. Heyer G, Hornstein OP, Handwerker HO. Skin reactions and itch sensation induced by epicutaneous histamine application in atopic dermatitis and controls. J Invest Dermatol 1989; 93:492-6.
35. Uehara M. Reduced histamine reaction in atopic dermatitis. Arch Dermatol 1982; 118:244-5.
36. Heyer G, Koppert W, Martus P et al. Histamine and cutaneous nociception: histamine-induced responses in patients with atopic eczema, psoriasis and urticaria. Acta Derm Venereol 1998; 78:123-6.
37. Heyer GR, Hornstein OP. Recent studies of cutaneous nociception in atopic and non-atopic subjects. J Dermatol 1999; 26:77-86.
38. Heyer G, Hornstein OP, Handwerker HO. Reactions to intradermally injected substance P and topically applied mustard oil in atopic dermatitis patients. Acta Derm Venereol 1991; 71:291-5.
39. Ishiuji Y, Coghill RC, Patel TS et al. Distinct patterns of brain activity evoked by histamine-induced itch reveal an association with itch intensity and disease severity in atopic dermatitis. Br J Dermatol 2009; 161:1072-80.
40. Schneider G, Stander S, Burgmer M et al. Significant differences in central imaging of histamine-induced itch between atopic dermatitis and healthy subjects. Eur J Pain 2008; 12:834-41.
41. Slater JW, Zechnich AD, Haxby DG. Second-generation antihistamines: a comparative review. Drugs 1999; 57:31-47.
42. Kelsay K. Management of sleep disturbance associated with atopic dermatitis. J Allergy Clin Immunol 2006; 118:198-201.
43. Lever R, Hefni A, Moffatt JD et al. Effect of tecastemizole on pulmonary and cutaneous allergic inflammatory responses. Clin Exp Allergy 2007; 37:909-17.
44. MacGlashan D Jr. Histamine: A mediator of inflammation. J Allergy Clin Immunol 2003; 112:S53-9.
45. Boone M, Lespagnard L, Renard N et al. Adhesion molecule profiles in atopic dermatitis vs. allergic contact dermatitis: pharmacological modulation by cetirizine. J Eur Acad Dermatol Venereol 2000; 14:263-6.
46. Zuberbier T, Henz BM. Use of cetirizine in dermatologic disorders. Ann Allergy Asthma Immunol 1999; 83:476-80.
47. Lippert U, Kruger-Krasagakes S, Moller A et al. Pharmacological modulation of IL-6 and IL-8 secretion by the H1-antagonist decarboethoxy-loratadine and dexamethasone by human mast and basophilic cell lines. Exp Dermatol 1995; 4:272-6.
48. Molet S, Gosset P, Lassalle P et al. Inhibitory activity of loratadine and descarboxyethoxyloratadine on histamine-induced activation of endothelial cells. Clin Exp Allergy 1997; 27:1167-74.
49. Rossbach K, Wendorff S, Sander K et al. Histamine H4 receptor antagonism reduces hapten-induced scratching behaviour but not inflammation. Exp Dermatol 2009; 18:57-63.
50. Dunford PJ, Williams KN, Desai PJ et al. Histamine H4 receptor antagonists are superior to traditional antihistamines in the attenuation of experimental pruritus. J Allergy Clin Immunol 2007; 119:176-83.
51. Suzuki K, Morokata T, Morihira K et al. A dual antagonist for chemokine CCR3 receptor and histamine H1 receptor. Eur J Pharmacol 2007; 563:224-32.
52. Amano T, Takeda T, Yano H et al. Olopatadine hydrochloride accelerates the recovery of skin barrier function in mice. Br J Dermatol 2007; 156:906-12.
53. Tamura T, Matsubara M, Amano T et al. Olopatadine ameliorates rat experimental cutaneous inflammation by improving skin barrier function. Pharmacology 2008; 81:118-26.
54. Diepgen TL. Early Treatment of the Atopic Child Study G. Long-term treatment with cetirizine of infants with atopic dermatitis: a multi-country, double-blind, randomized, placebo-controlled trial (the ETAC trial) over 18 months. Pediatr Allergy Immunol 2002; 13:278-86.

55. Kawashima M, Tango T, Noguchi T et al. Addition of fexofenadine to a topical corticosteroid reduces the pruritus associated with atopic dermatitis in a 1-week randomized, multicentre, double-blind, placebo-controlled, parallel-group study. Br J Dermatol 2003; 148:1212-21.
56. Berth-Jones J, Graham-Brown RA. Failure of terfenadine in relieving the pruritus of atopic dermatitis. Br J Dermatol 1989; 121:635-7.
57. Wahlgren CF, Hagermark O, Bergstrom R. The antipruritic effect of a sedative and a nonsedative antihistamine in atopic dermatitis. Br J Dermatol 1990; 122:545-51.
58. Hannuksela M, Kalimo K, Lammintausta K et al. Dose ranging study: cetirizine in the treatment of atopic dermatitis in adults. Ann Allergy 1993; 70:127-33.
59. Kawakami T, Kaminishi K, Soma Y et al. Oral antihistamine therapy influences plasma tryptase levels in adult atopic dermatitis. J Dermatol Sci 2006; 43:127-34.
60. Maurer M, Worm M, Zuberbier T. Antihistamines in atopic dermatitis. In: Reitamo S, Luger T, Steinhoff M, eds. Textbook of Atopic Dermatitis. London: Informa UK Ltd, 2007.
61. Vuurman EF, van Veggel LM, Uiterwijk MM et al. Seasonal allergic rhinitis and antihistamine effects on childresn's learning. Ann Alergy 1993; 71:121-6.
62. Boyle J, Eriksson M, Stanley M et al. Allergy medication in Japanese volunteers: treatment effect of single doses on nocturnal sleep architecture and next day residual effects. Curr Med Res Opin 2006; 22:1343-51.
63. Williams DH. Skin temperature reaction to histamine in atopic dermatitis (disseminated neurodermatitis). J Invest Dermatol 1938; 1:119-29.

Chapter 9

Histamine, Immune Cells and Autoimmunity

Elke Schneider, Maria Leite-de-Moraes and Michel Dy*

Abstract

Histamine is one of the most versatile biogenic amines with multiple roles during the immune response and in allergic disorders. With four distinct G protein-coupled receptors (H_1R, H_2R, H_3R and H_4R), intracellular histamine binding sites (most likely members of the cytochrome P450 family) as well as a membrane transporter (Organic Cation Transporter; OCT3) expressed in various immunocompetent cells, it can entertain a complex network of interactions. These signaling pathways are expressed differentially, depending on the stage of differentiation or activation of target cells, thus adding a further degree of complexity to the system. For this reason, published data are sometimes conflicting and varying according to the particular cell type or responses analyzed and the experimental approaches used. On the other hand, histamine is generated by several cells during the immune response, not only through release of intracellular stores in mast cells or basophils in response to IgE-dependent or -independent stimuli, but also through neosynthesis catalyzed by histidine decarboxylase (HDC) in a number of hematopoietic cells that secrete the amine immediately without prior storage. These features enable histamine to tune the fine balance between immunity and tolerance by affecting dendritic cells, immunoregulatory cells, T-cell polarization and cytokine production, making the way for new pharmacological strategies to control immune reactivity during immune disorders, such as autoimmunity.

Introduction

Histamine (2-(imidazol-4-yl) ethylamine) was discovered in 1910 by Sir Henry Dale,[1] due to its ability to constrict guinea-pig ileum. At present, it is considered the biogenic monoamine with the broadest spectrum of activities in various physiological and pathological situations. Thus, it performs neurotransmitter functions in the central nervous system, regulates peripheral vasoactivity as well as acid secretion in the stomach and modulates immune responses, inflammation and hematopoiesis. These effects are mediated through four distinct histamine receptors (H_1R, H_2R, H_3R and H_4R), which are heptahelical, G-protein-coupled molecules expressed either ubiquitously (H_1R and H_2R) or predominant in particular tissues (H_3R in the brain and H_4R in the hematopoietic system). The multiple activities of histamine and its receptors have been extensively reviewed.[2-8]

Histamine is synthesized by a unique enzyme, histidine decarboxylase (HDC) (EC.4.1.1.22) that requires pyridoxal-5-phosphate as a cofactor. The HDC gene is located on chromosome 15 in humans and chromosome 2 in mice and its expression is controlled by various lineage-specific transcription factors.[2] Recently, several findings have shed a new light on the contribution of histamine to the regulation of the immune response, namely 1) cloning of a H_4R expressed in hematopoietic

*Corresponding Author: Michel Dy—Université Paris Descartes, Faculté de Médecine, CNRS UMR8147, Hôpital Necker Paris, France. Email: dy@necker.fr

Histamine in Inflammation, edited by Robin L. Thurmond.

cells, 2) demonstration of histamine synthesis in immuno-competent cells other than mast cells or basophils, 3) identification of OCT3 as a transporter through which intracellular histamine levels can be increased to inhibit basophil functions and 4) evidence for a histamine-cytokine connection.[2] These properties enable histamine to modulate the fine balance that prevents the rupture of immune tolerance toward various tissue autoantigens leading to autoimmunity. We will discuss this issue later in this chapter, once we have dealt with the cellular sources of histamine and its effect on target cells involved in the immune response.

Histamine and Immune Cells

Basophils and mast cells have long been considered the unique source of histamine among the cells of the immune system. They remain the most proficient producers of this biogenic amine, since they can store and release in response to IgE-dependent or independent stimuli. However, it has been established that a number of other immuno-competent cells can express high levels of inducible HDC activity and secrete the newly synthesized histamine immediately rather than storing it in specific granules. This property is shared by dendritic cells, neutrophils and monocytes/macrophages and lymphoid cells that generate histamine in response to various stimuli, making it available in the microenvironment, ready to modulate the biological activities of other immune cells and hence the orientation of the immune response.

Histamine and Dendritic Cells

Dendritic cells (DCs) are professional antigen-presenting cells of lymphoid or myeloid origin, present in a variety of tissues. Immature DCs are activated by pathogens and cytokines that promote their final maturation into the DC1 or DC2 phenotype and their migration into lymphoid organs where they activate resting T-lymphocytes and produce cytokines that determine the differentiation of $CD4^+$ T cells into different helper subsets. DCs express H_1R, H_2R and H_4R, while H_3R expression is low or undetectable.[9-13] Histamine modulates their typical functions, such as chemotaxis,[11,12] antigen uptake and cross-presentation,[9] cytokine and chemokine production[14-16] as well as their ability to drive $CD4^+$ T-cell differentiation[14,16] by targeting one or several of its receptors, depending on their respective surface expression. In this context, several investigators have examined the effect of histamine on the capacity of DCs to promote the transformation of naïve $CD4^+$ T cells into Th1, Th2 or Th17 cells. They established that the amine inhibits IL-12 p70 and increases IL-10 production through H_1R, H_2R and/or H_4R activation, thus favoring the development of Th2 cells.[12,14,16,17] Histamine induces chemotaxis of human immature DCs by targeting H_1R and H_2R,[15] while chemotaxis of murine bone marrow-derived DCs is enhanced via the H_4R, as assessed in vitro as well as in a skin model in vivo.[11] This receptor is also implicated in the enhancement of the cross-presentation of antigen by MHC-class I molecules induced by exposure of immature DCs to histamine,[9] while its positive effect on antigen uptake and endocytosis is mediated through the H_2R subtype.[9] These data suggest that histamine can enhance the ability of extracellular antigens to activate $CD8^+$ T-cell-mediated responses by targeting DCs. Its effect is restricted to soluble antigens, while particulate antigen cross-presentation or uptake by dendritic cells is not affected.[18] Not only does histamine influence DC polarization to skew the differentiation of naive T cells toward a Th2 profile, but it also enhances Th2 cell recruitment by inducing Th2-attracting chemokines (CCL17 and CCL2), while inhibiting their Th1 counterpart (CXCL10).[18]

Plasmacytoid DCs (pDCs) constitute another subset of professional antigen-presenting cells and are a major source of IFNα. Similarly to what happens in myeloid DCs, histamine modulates their cytokine production through H_2R. Indeed, the presence of histamine during stimulation of pDCs by live flu virus or CpG oligodeoxynucleotides markedly decreases their IFNα and TNFα production.[19] This may explain why viral infections in atopic children are associated with low levels of Type I IFN. In striking contrast with functional H_1R and H_2R expression by myeloid DCs and dermal dendritic cells, Langerhans cells lack both receptors, probably because their expression is inhibited by TGFß1, which is required for the differentiation of these cells.[20]

Interestingly, previous studies have demonstrated that DCs themselves can produce histamine, which could in turn modulate the expression of DC markers in an autocrine or paracrine manner.[21] It is tempting to speculate that during inflammatory processes DCs can produce sufficient amounts of histamine to act similarly, since it has been reported that histamine production by DCs is increased under such circumstances.[22] In support of such a contribution to antigen presentation and regulation of Th1/Th2 $CD4^+$ T-cell differentiation, it has been described that the antigen-presenting capacity and the cytokine production profile are altered in spleen DCs from HDC-deficient mice, leading to preferential Th1 development.[23]

Histamine and T Cells

It is currently acknowledged that histamine can influence T helper cell differentiation by targeting DCs. This notion is supported by the decreased allergic airway inflammation in an allergic asthma model carried out in mice in which the H_4R was either disrupted or blocked by a specific antagonist.[22] These data are reminiscent of a similar effect described in mice lacking HDC or injected with histamine-binding proteins.[24-26]

On the other hand, histamine receptors are also expressed by T cells, which respond directly to the amine. Indeed, Th1 cells display predominantly the H_1R, through which histamine enhances their typical functions, while the H_2R that mediates the inhibitory effect of histamine on Th2 as well as on Th1 cells is preferentially associated with the Th2 subset.[27,28] In agreement with these data, H_1R-deficient mice produce low levels of IFNγ together with high amounts of Th2-derived cytokines, while both Th1- and Th2-type cytokine synthesis is increased in their H_2R-deficient counterpart. Although IL-17 production is diminished during the asthmatic response in H_4R-deficient mice,[24] no formal demonstration of the effect of histamine on Th17 cell differentiation has been provided as yet. Conversely, it has been shown that histamine does not affect Th17 cell differentiation in a model involving mast cells.[29]

$CD8^+$ T cells are also sensitive to histamine as demonstrated by their increased IL-16 production in response to H_2R or H_4R engagement[30] and by their reduced IFNγ production in H_1R- or H_2R-deficient mice.[31]

Endogenous production of histamine by $CD4^+$ and $CD8^+$ T cells has been described following mitogen stimulation.[31-32] Although HDC has been detected in the Jurkat cell line,[33] normal T cells need to be purified more thoroughly to confirm their histamine production since this could easily be generated by a few contaminating basophils or basophil precursors (less than 1%). This explanation is particularly likely in view of the authors' claim that IL-3 and GM-CSF, two cytokines well known for their effect on this lineage, increase histamine production in the lymphocyte preparation, even in the absence of mitogen.[32] Indeed, it has been documented long ago that splenic nonT non-B cells,[34] presently identified as basophils[35,36] increase their histamine synthesis in response to IL-3 and GM-CSF, both produced by ConA-stimulated lymphocytes.[37-39]

Histamine and Immunoregulatory T Cells

The immunosuppressive functions of histamine have been known for a long time and were initially ascribed to its ability to induce IL-10 production, a strong immunosuppressive or immunoregulatory cytokine. As mentioned above, histamine targets dendritic cells or Th2 cells to increase their production of IL-10, which can in turn enhance the suppressive effect of TGFβ on T cells.[40] More recently, the effect of histamine on immunoregulatory T cells, such as $CD4^+CD25^+Foxp3^+$ (Treg) and NKT cells, has been investigated in an allergic asthma model. This study established that histamine acted as a chemoattractant of T cells by activating their H_1R or H_4R. However, those recruited through the H_1R were mainstream T-lymphocytes, whereas those targeted via the H_4R belonged mostly to the $CD4^+CD25^+Foxp3^+$ T-cell subset that suppressed autologous T-cell proliferation in an IL-10-independent fashion. These regulatory T cells accumulated in the lung following instillation of H_4R agonist and might be responsible for the inhibition of allergic asthma in this model.[41] However, the apparent discrepancy between this result and a similar alleviation of disease syndromes reported after treatment with a specific H_4R antagonist[25] needs to be clarified, even though the route of administration differs between the two protocols (local for the agonist versus systemic for the antagonist).

iNKT cells, another immunoregulatory T-cell subset, constitute a distinctive population of mature T-lymphocytes positively selected by the nonpolymorphic MHC class-I-like molecule, CD1d. They co-express a highly restricted T-cell receptor (TCR) repertoire, composed of a single invariant Vα14Jα18 chain in mice and a Vα24Jα18 chain in humans, preferentially paired with a limited TCR Vβ chain repertoire that specifically recognizes glycolipids. iNKT cells are implicated in the control of several immune responses, most likely because of their capacity to promptly produce cytokines, such as IL-4 and IFN-γ.[42-44] Histamine has been shown to target this cell population by modulating their cytokine production. Indeed, iNKT cells from HDC-deficient mice generate less cytokines in response to their specific ligand αGalCer than their wild-type counterpart. Administration of histamine restores a normal production, through H_4R engagement. Indeed, this conclusion is supported by the fact that the functional recovery no longer occurs when the receptors are blocked by a specific antagonist and a similar deficit in iNKT-cell-derived cytokine production is found in H_4R- and HDC-deficient mice.[45] Although the exact mechanism through which histamine exerts this positive effect has not been elucidated, these data underscore once again the importance of mutual interactions between histamine and cytokines, whatever the target cells.[2,46]

Histamine and B Cells

Anti-IgM-induced B cell proliferation in mice is increased in the presence of histamine and diminished in H_1R-deficient mice, suggesting that H_1R activation can amplify B cell receptor signaling. Concerning the antibody response to T-cell-dependent antigens, two different results have been reported, namely an increase of ovalbumin-specific IgE and IgG1 antibody production in H_1R-deficient mice and a decreased IgE and IgG3 production in H_2R-deficient mice. In the latter, ovalbumin-specific IgE levels dropped, in spite of the enhanced IL-4 and IL-13 production, because of the high inhibitory concentration of IFNγ. This finding supports the idea that H_1R and Th1 responses prevail over humoral responses.[29,47]

Histamine and Monocytes/Macrophages

Histamine decreases p40 and p70 IL-12 and increases IL-10 production through the H_2R in lipopolysaccharide (LPS)-stimulated whole blood cells or purified monocytes.[48,49] These data are reminiscent of the work of Rocklin et al. who demonstrated the presence several years ago of a histamine-induced suppressor T-cell factor derived from monocytes,[50] which, in the light of the present data, could be identical with IL-10. Histamine also inhibits LPS-induced TNFα production by human peripheral blood monocytes via its H_2R.[51] Conversely, it fails to prevent LPS-induced upregulation of TNFα expression in macrophages or even increases its secretion by modulating the TNFα-converting enzyme (TACE) via H_1R.[52] A distinctive effect of histamine on monocytes and macrophages is also observed in terms of Ca^{++} influx and IL-8 production in response to H_1R stimulation, which takes place only in macrophages.[53] Furthermore, decreased lectin-like oxidized low-density receptor-1 (LOX-1) gene expression associated with upregulation of monocyte-chemoattractant protein-1 (CCL2) and its receptor CCR2 via H_2R engagement occurs in monocytes but not in macrophages.[54,55] This differential modulation is explained by a switch in histamine receptor expression from H_2R to H_1R during maturation of monocytes into macrophages.[52,53] In apparent contradiction, it has recently been shown that CCL2 synthesis and secretion by monocytes is downregulated by histamine through H_4R.[56]

The decreased TNFα production by LPS-induced monocytes in the presence of histamine might result from its ability to reduce the surface expression of CD14, but not TLR4.[57] The modulation of CD14 probably occurs through posttranscriptional events, since mRNA levels remained unchanged. However, histamine does not downregulate this surface marker during GM-CSF- and IL-4-induced differentiation of monocytes into dendritic cells that continue to express CD14 but not CD1a.[58] It is also noteworthy that histamine diminishes IL-18- induced IFNγ, TNFα and IL-12 production by human PBMC. IL-18 exerts this effect through upregulation of ICAM on monocytes, which is prevented by histamine via the H_2R.[59] In addition, it has also been demonstrated that histamine increases the lifespan of monocytes by protecting them against apoptosis in response to CD95/Fas ligation, dexamethasone or serum deprivation. These effects are explained by upregulation of Bcl-2 and Mcl-1 and inhibition of caspase 3 activation and could be partially mediated through histamine-induced IL-10 production.[60]

HDC expression increased during maturation of monocytes into macrophages,[55] in agreement with other reports on histamine synthesis in the differentiated population.[54,61] In addition, it has been shown that mouse peritoneal macrophages as well as the macrophage cell line, RAW264.7 can take up histamine and release it when its extracellular concentrations drop.[62] Whatever the exact mechanism, these data suggest that histamine originating from macrophages could contribute to their deleterious effects during inflammatory pathologies, such as in atherosclerosis.[55]

Histamine and Basophils/Mast Cells

Mast cells and basophils compose the main population of cells in which histamine can be stored to be promptly liberated upon stimulation. Mast cells reside in various tissues of the organism, conversely to basophils, which represent the mobile pool of the amine. Both cells derive from $CD34^+$ hematopoietic stem cells. Mast cells leave the bone marrow as immature precursors and complete their differentiation in peripheral tissues. Conversely, basophils enter the circulation only when they have achieved full maturation in the bone marrow. It is generally accepted that mast cells and basophils represent distinct cell lineages derived from different progenitors. However, some data argue in favor of a mast cell/basophil progenitor, such as the expression of a common antigen recognized by the antibody 97A6, shared by mature basophils and mast cells, as well as their precursors[63] and the identification of cells with metachromatic granules combining the features of both basophils (blood location, segmented nuclei and expression of Bsp1, a basophil specific antigen) and mast cells (c-kit, tryptase and chymase expression) in the peripheral blood of patients with asthma, allergy and allergic drug reactions.[64] Mast cells and basophils are regarded as key effector cells in IgE-associated immediate hypersensitivity reactions and allergic disorders, while basophils though described over a century ago, remain enigmatic as to their physiological functions. However, recent data suggest that they may play an important role during helminth infections and are more efficient than mast cells in producing IL-4 together with histamine, which both facilitate Th2 differentiation.[65] Mast cells and basophils share the expression of FcεRI, a tetramer composed of one α, β and two γ chains ($\alpha\beta\gamma2$). Cross-linking of FcεRI-bound IgE with antigen, initiates degranulation with subsequent release of stored mediators, such as histamine, de novo synthesis of pro-inflammatory lipid mediators and production of cytokines and chemokines. In these conditions, the amount of histamine liberated into the microenvironment may reach millimolar levels. This process is enhanced by high concentrations of IgE, which upregulate membrane FcεRI expression. In addition, recent data indicate that monomeric IgE can increase survival of mast cells without cross-linking, by rendering them resistant to apoptosis. This type of stimulation is efficient enough to induce cytokine production and increased HDC activity through a signaling pathway distinct from the one triggered by antigen-induced FcεRI cross-linking.[66]

Although basophils and mast cells are primarily a source of histamine, they also express histamine receptors (H_1R, H_2R and H_4R) and transporters (OCT3) and could therefore be targeted by the amine in an autocrine or paracrine manner. For instance, the H_4R seems to be involved in the control of mast cell chemotaxis since this biological activity is induced by receptor engagement in vitro and results in a change of tissue localization in vivo.[67,68] Histamine also synergizes with chemoattractants, such as CXCL12 by targeting the H_4R on mast cell precursors.[69] Histamine does not seem to affect degranulation in either cell. However, in basophils, histamine exerts a negative control on its own synthesis and that of associated cytokines (IL-4, IL-6 and IL-13). This effect is not mediated through classical receptors, but results from increased intracytosolic histamine levels under the control of the organic cation transporter, OCT3. When intracellular histamine attains a critical level, it inhibits the transcription of HDC and cytokine genes[70] by a mechanism not clearly identified as yet, but most likely related to molecules of the CYP450 family[71,72] (Fig. 1).

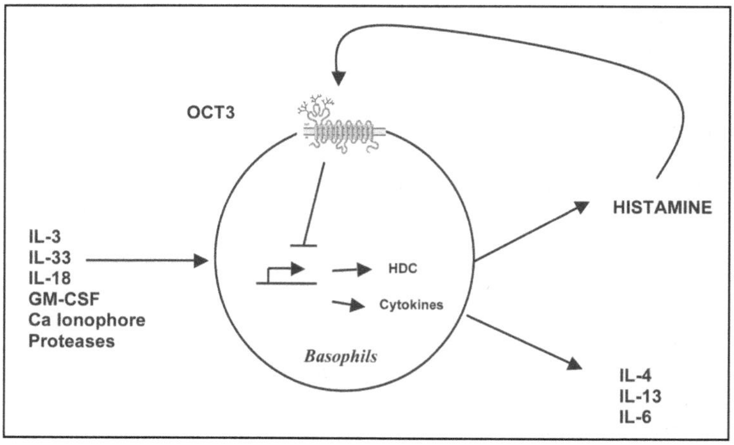

Figure 1. Basophils: a typical cell at the crossroad between cytokines and histamine. IL-3 and some other cytokines increase histamine synthesis as well as pro-Th2 cytokine production by basophils. When extracellular levels of histamine are high, histamine is taken up by OCT3 and inhibits its own synthesis and those of associated cytokines.

Histamine and Eosinophils/Neutrophils

As stated above for mast cells, histamine is also a potent chemoattractant for eosinophils via H_4R activation. Likewise, changes in eosinophil shape and increases in expression of adhesion molecules like CD11b/CD18 (Mac1) and CD54 (ICAM-1) appear to be mediated through this receptor.[73-75] It has also been claimed that the chemokine (LEC)/CCL16, which is expressed in the liver, targets the H_4R, causing human or mouse eosinophil migration.[76] This finding suggests an intriguing functional similarity between chemokines and H_4R that needs to be confirmed. Using a protein array to evaluate cytokine production by human eosinophils, it has been shown that levoceterizine (H_1R antagonist) inhibits IL-1, IL-7 and SCF production promoted by stimulation with LPS.[77] At relatively high doses (10-100 μM) histamine can also counteract the effect of IL-5 on the survival of human eosinophils by inducing their apoptosis. This effect occurs through an unknown mechanism that does not involve classical receptors,[78] but might be analogous to the cAMP-dependent-apoptotic pathway induced by some H_4R agonists, which inhibit antigen-specific human T-cell responses through an H_4R-independent pathway.[79]

Histamine is also involved in chemotaxis of neutrophils, as shown during their mast cell-dependent recruitment induced by zymosan in vivo[80] and trinitrobenzene sulphonic acid-provoked acute colitis.[81] This effect is mediated through the H_4R, which is likewise responsible for the decrease in bone marrow neutrophils following injection of histamine.[80] In addition to being a target, neutrophils are also a source of histamine, as evidenced in a casein-induced peritonitis model where HDC has been localized on the intracytosolic face of the membrane granules[82] as well as during mycoplasma pneumonia.[83] In this latter model, mycoplasma has been shown to stimulate naive neutrophils directly to synthesize histamine by strongly upregulating HDC mRNA expression. Further investigations will be required to elucidate the nature of signals exchanged between mycoplasma and neutrophils that lead to such an increase in HDC mRNA expression.

Histamine and Autoimmunity

As discussed above, histamine controls accessibility to sites of inflammation by modulating vasopermeability and adhesion molecule expression. In addition, it exerts a chemotactic effect on various cell types, on its own or in synergy with classical chemoattractants. It also targets DCs and Th1/Th2 cells directly, mainly by modulating their cytokine profile, thus establishing

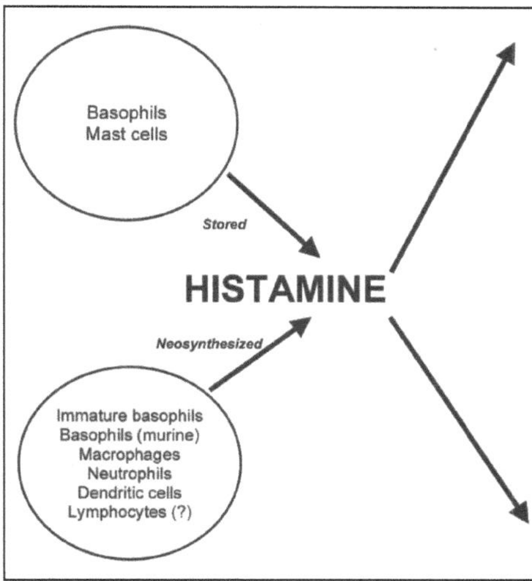

Histamine targets	Cytokines	Receptors	Effect	Ref.
Dendritic cells	IL-12	H1, H2, H4	decrease	12, 14, 16
	IL-10	H2	increase	16,17
	IFNα	H2	decrease	20
	TNFα	H2	decrease	20
TH1 cells	IFNγ	H1	increase	28
TH2 cells	IL-4	H2	decrease	28
CD8+T cells	IL-16	H4	increase	30
iNKT cells	IL-4	H4	increase	44
	IFNγ	H4	increase	44
Monocytes	IL-12	H2	decrease	48,49
	IL-10	H2	increase	48
	TNFα	H2	decrease	51
	CCL2	H2, H4	?	54 ,56
Macrophages	TNFα	H1	increase	52
	IL-8	H1	increase	53
Basophils	IL-4	intra	decrease	70
	IL-6	intra	decrease	70
	IL-13	intra	decrease	70

Figure 2. Histamine-cytokine connection in immune cells.

the histamine/cytokine network illustrated in Figure 2. These multiple biological activities are consistent with a major regulatory function of histamine during the immune response and in the emergence of pathologies resulting from immune disorders such as autoimmunity. In favor of a potential immunointervention, increased production of histamine occurs during diseases such as multiple sclerosis,[84] psoriasis,[85] Crohn's disease and ulcerative colitis[86,87] in which histamine metabolite levels are correlated with disease severity. In the same line of evidence, the increase of mast cell numbers associated with autoimmune diseases argues in favor of the contribution of histamine during onset or effector phase of the pathology, even though it is quite evident that mast cells participate in various other ways in the autoimmune reaction, as summarized in a recent review.[88] For instance, mast cells are found in the brain of patients with multiple sclerosis associated with demyelinated plaques,[89-91] in the synovial fluid of rheumatoid arthritis patients,[93-95] in the salivary gland of individuals suffering from Sjogren's syndrome,[96] in the vicinity of peripheral nerves during experimental neuritis and so on. Basophils, as the second most potent source of histamine, could also participate in these processes, even though there is no reliable evidence for their role in autoimmunity as yet, possibly because these rare cells are difficult to identify. The availability of new specific surface markers and murine models in which basophils have been implicated during the polarization of the immune response,[97] will certainly lead to the reappraisal of their role in the near future. In addition, histamine receptor-bearing cells have been recovered from sites of autoimmune aggression, as exemplified by the presence of infiltrating H_1R and H_2R-positive cells in the brain in experimental autoimmune encephalomyelitis (EAE) models and in synovial fluids of rheumatoid arthritis patients, which contain fibroblasts and macrophages that express H_4R.

Histamine and Experimental Autoimmune Encephalomyelitis (EAE)

The large majority of studies looking for a possible implication of histamine during autoimmune diseases has been performed in the murine EAE model, which is most closely related to human multiple sclerosis (MS). Mice are immunized with myelin peptide in the presence of complete Freund adjuvant to generate the autoimmune disease in which Th1[98] and/or Th17 cells are most likely involved. Indeed, IFNγ- or IFNγ receptor-deficient mice express a more severe EAE,[99,100] while IL-23-deficient mice are relatively protected.[101] The etiology of the disease, characterized by myelin destruction in the CNS, is not entirely clear. In MS or EAE, one of the earliest events of disease onset is an increase in blood barrier permeability that allows inflammatory and immune cells to infiltrate the murine CNS, a step in which the vasoactive properties of histamine could play a major role. In support of this assumption, mice with disrupted H_3R expression develop a more severe disease and neuroinflammation, due to deficient neurogenic control of cerebrovascular tone as well as increased chemokine production.[102] Moreover, histamine can enhance disease progression through the H_1R, as shown by reduced pathological alterations after receptor blockade[98] as well as a significant delay in EAE onset in H_1R-deficient mice, associated with a decrease in the severity of clinical symptoms.[103] By contrast, it has been reported that H_2R activation by the agonist dimaprit has a beneficial effect, alleviating the typical hallmarks of disease.[104] These data, together with the fact that during EAE inflammatory and immune cells bearing H_1R or H_2R infiltrate the CNS, argue strongly in favor of the contribution of histamine to this pathology. This notion is strengthened by the exacerbation of MOG 35-55-induced chronic EAE in histamine-free mice. Indeed, CNS inflammatory infiltrates that develop in brain parenchyma of these mutants are more diffuse and contain more eosinophils and polymorphonuclear leukocytes than their wild-type counterpart. Furthermore, T cells from HDC-deficient mice produce more IFNγ, TNFα and MCP-1 in response to autoantigens, suggesting that the overall effect of histamine might be protective, limiting the CNS immune damage.[105]

Taken together these data are somewhat contradictory. On the one hand, the reduced pathogenicity observed in H_1R KO mice might be explained by decreased inflammatory cell infiltration in the CNS because the vasopermeability at the BBB (Blood-Brain Barrier) is no longer increased via this receptor subtype. In accordance with this finding the disease is exacerbated in H_3R KO mice whose histamine synthesis is no longer regulated by the H_3R, leading to increased vasopermeability, which facilitates the infiltration of immunocompetent cells in the CNS. These data are reminiscent of similar chain of events in cerebral malaria.[106] On the other hand, the question arises why the lack of histamine in HDC KO mice aggravates the disease. To account for this result the control of the vascular tone by histamine might be considered solely as a means of enhancing the pathology by favoring the immune cell infiltration in the CNS, while more pathogenic T_H1 cells are generated and migrate together with other inflammatory cells to the CNS in a histamine-independent fashion. This hypothesis is supported by the observation that IFN-γ production by T cells from HDC KO mice is increased. Moreover, the activity of infiltrating cells could be increased in the absence of histamine, which is known for downregulating leukocyte functions such as production of oxygen radicals, leukotrienes and cytokines.

In agreement with the regulatory functions of histamine in autoimmune diseases, the *Bordetella pertussis* toxin-induced histamine sensitization (Bphs) gene that controls the susceptibility to EAE and experimental allergic orchitis, has been identified as the H_1R.[103] Signaling through this receptor is important during early activation of $CD4^+$ T cells since it is required for their TCR-mediated p38 MAPK activation and optimal IFNγ production.[107] Indeed, structural polymorphism (L263P, M313V and S331P) in the third intracellular loop of the murine H_1R regulates T-cell cytokine production and thereby controls disease susceptibility. The PVP haplotype is associated with increased susceptibility (H_1R^s), while the LMS counterpart develops a less severe disease (H_1R^r). Mechanistically, polymorphism alters H_1R surface expression; the H_1R^r allele being retained within the endoplasmic reticulum of T cells, thus modifying their immune functions and autoimmune disease susceptibility.[108]

Histamine and Autoimmune Chronic Urticaria (CU)

Chronic urticaria is a common disease characterized by recurrent, transitory and itchy wheals for more than six weeks that may severely worsen the quality of life. No precise pathogenesis has been established so far for all cases of CU, although a serologic component has been identified in many cases of autoimmune origin.[109] Indeed, in around 50% of CU patients circulating antibodies directed against the high affinity receptor for IgE (FcεRI) and more rarely against IgE have been detected.[110] These autoantibodies are responsible for in vitro histamine release from basophils or mast cells and explain the in vivo wheal-and-flare response observed following intradermal injection of autologous serum (autologous serum skin test, ASST). In addition, the binding of these autoantibodies to receptors triggers the activation of the complement cascade and production of anaphylatoxins such as C5a that synergizes in turn with FcεRI autoantibodies to enhance histamine release from mast cells.[111] The autoimmune origin of CU is also supported by its association with thyroid autoimmune disease in some patients.[112] In a limited number of cases, skin biopsies with infiltrating $CD4^+$ T cells contained some $CD4^+CD25^+$ Treg cells characterized by an uncommon decrease in their ability to inhibit $CD4^+CD25^-$ proliferation in response to mitogen.[113]

Circulating autoantibodies directed against high-affinity IgE receptors were also found in some patients suffering from asthma, suggesting that autoimmunity may contribute to intrinsic asthma pathogenesis.[114]

Histamine and Rheumatoid Arthritis (RA)

Rheumatoid arthritis is defined as an autoimmune disease with chronic inflammation of the synovium that leads to the destruction of bone and articular cartilage. Histamine is found both in diseased synovium and joint fluid[115-118] and originates either from an increased number of activated and degranulated mast cells at the inflammatory sites or from neosynthesis by chondrocytes of osteoarthritic cartilage.[119,120] Based on these data and the fact that histamine receptor-bearing cells are present in the synovium,[121,122] it has been suggested that histamine could increase inflammation during RA.[120,123] However, histamine injected in mouse knee joints does not induce any signs of synovitis on its own. In addition, even in combination with HMGB1 or peptidoglycans, histamine injection does not modify the inflammatory effect of these molecules. The fact that mast cell membrane stabilization does not alter in vivo inflammatory responses supports the idea that histamine is not responsible for this process. Lastly, a recent study shows that histamine levels both in synovial fluids and in sera of patients suffering from RA are significantly lower than in healthy individuals and anti-TNFα treatment of RA patients restores normal histamine levels.[124] Taken together, these data does not fit with previous data and argue in favor an anti-inflammatory rather than a pro-inflammatory role of histamine during RA.

Histamine and Experimental Autoimmune Myocarditis

Experimental autoimmune myocarditis represents another model in which a possible implication of histamine has been evaluated. The development of this pathology is associated with H_1R expression in the myocardium, which does not occur in healthy individuals.[125-127] It might be hypothesized that histamine provided by infiltrating cardiac mast cells impairs cardiomyocyte functions via H_1R activation, as suggested by the improvement of viral myocarditis following treatment with H_1R antagonists.[128]

Conclusion

The regulatory functions of histamine during the immune response are widely documented. However, the complexity of interactions between immune cells through a variety of receptors and other binding sites has engendered some conflicting data. They are probably explained by the relative selectivity of histamine receptor agonists and antagonists used in these studies, depending on their concentrations and the identity of target cells. Moreover, the contribution of other sites of interaction, such as membrane transporters like OCT3 or intracellular receptors has certainly been underestimated in the overall effect of histamine. The most recent discovery of the H_4R and its predominant expression in hematopoietic and immunocompetent cells has led to a reappraisal of the role of histamine during the immune response and provided a new pharmacological target with potential therapeutic applications. Although interesting data have already been obtained in some models of autoimmune diseases, the appreciation of the influence of histamine on the equilibrium between immunity and tolerance and its complex network of interactions is far from complete. Differences between species, routes of administration of histamine response modifiers and the like, are probably responsible for the confusing picture obtained so far. This needs to be put into better focus in order to evaluate the impact of future therapeutic strategies. Furthermore the involvement of histamine during self-recognition has yet to be addressed, although recent data implicate histamine during T-cell tolerance to high dose bee venom exposure in beekeepers that is caused by an in vivo switch from venom-specific Th2 to IL-10-secreting Treg cells via H_2R activation.[129]

Finally, the peculiar interaction between histamine and cytokines raises hope for new pharmaceutical developments of histamine-related molecules acting on the inflammatory axis of autoimmune diseases. For example, the demonstration that H_4R antagonists affect TNFα production in a model of colitis could lead to new treatments of autoimmune diseases that are so far based mainly on anti-TNFα therapies.

References

1. Barger G, Dale HM. The presence in ergot and physiological activity of B-iminazoylethylamine. J Physiol Paris 1910; 40:38-40.
2. Dy M, Schneider E. Histamine-cytokine connection in immunity and hematopoiesis. Cytokine and Growth Factor Rev 2004; 15:393-410.
3. Huang JF, Thurmond RL. The new biology of histamine receptors. Current Allergy and Asthma Reports 2008; 8:21-27.
4. Zhang M, Thurmond RL, Dunford PJ. The histamine H4 receptor: a novel modulator of inflammatory and immune disorders. Pharmacol Therapeut 2007; 113:594-606.
5. Jutel M, Blaser K, Akdis CA. Histamine in allergic inflammation and immune modulation. Int Arch Allergy Immunol 2005; 137:82-92.
6. Schneider E, Rolli-Derkinderen M, Arock M et al. Trends in histamine research: new functions during immune responses and hematopoiesis. Trends Immunol 2002; 23:255-263.
7. Jutel M, Blaser K, Akdis CA. The role of histamine in regulation of immune responses. In: Crameri R, ed. Molecular Aspects of Allergy and Asthma.: Allergy and asthma I modern society: A scientific approach. Chem Immunol Allergy. Basel: Karger, 2006:174-187.
8. Tanaka S, Ichikawa A. Recent advances in molecular pharmacology of the histamine systems: immune regulatory roles of histamine produced by leukocytes. J Pharmacol Sci 2006; 101:19-23.
9. Amaral MM, Davio C, Ceballos A et al. Histamine improves antigen uptake and cross-presentation by dendritic cells. J Immunol 2007; 179:3425-3433.
10. Damaj BB, Becerra CB, Esber HJ et al. Functional expression of H4 histamine receptor in human natural killer cells, monocytes and dendritic cells. J Immunol 2007; 179:7907-7915.
11. Baumer W, Wendorff S, Gutzmer R et al. Histamine H4 receptors modulate dendritic cell migration through skin—immunomodulatory role of histamine. Allergy 2008: 63:1387-1394.
12. Gutzmer R, Diestel C, Mommert S et al. Histamine H4 receptor stimulation suppresses IL-12p70 production and mediates chemotaxis in human monocyte-derived dendritic cells. J Immunol 2005; 174:5224-5232.
13. Dijkstra D, Stark H, Chazot PL et al. Human inflammatory dendritic epidermal cells express a functional histamine 4 receptor. J Invest Dermatol 2008; 128:1696-1703.
14. Caron G, Delneste Y, Roelandts E et al. Histamine polarizes human dendritic cells into Th2 cell-promoting effector dendritic cells. J Immunol 2001; 167:3682-3686.
15. Caron G, Delneste Y, Roelandts E et al. Histamine induces CD86 expression and chemokine production by human immature dendritic cells. J Immunol 2001; 166:6000-6006.
16. Mazzoni A, Young HA, Spitzer JH et al. Histamine regulates cytokine production in maturing dendritic cells, resulting in altered T-cell polarization. J Clin Invest 2001; 108:1865-1873.
17. Ozna N, Elliott K, Khan MM. Regulation of interleukin-10 secretion by histamine in TH2 cells and splenocytes. Int Immunopharmacol 2001; 1:85-96.
18. McIlroy A, Caron G, Blanchard S et al. Histamine and prostaglandin E_2 up-regulate the production of Th2-attracting chemokines (CCL17 and CCL22) and down-regulate IFN-γ-induced CXCL10 production by immature human dendritic cells. Immunology 2006; 117:507-516.
19. Mazzoni A, Leifer CA, Mullen GED et al. Cutting edge: histamine inhibits IFN-α release from plasmacytoid dendritic cells. J Immunol 2003; 170:2269-2273.
20. Ohtani T, Aiba S, Mizuashi M et al. H1 and H2 histamine receptors are absent on Langerhans cells and present on dermal dendritic cells. J Invest Dermatol 2003; 121:1073-1079.
21. Szeberenyi JB, Pallinger E, Zsinko M et al. Inhibition of endogenously synthesized histamine disturbs in vitro human dendritic cell differentiation. Immunol Lett 2001; 76:175-182.
22. Dunford PJ, O'Donnell N, Riley JP et al. The histamine H4 receptor mediates allergic airway inflammation by regulating the activation of $CD4^+$ T-cells. J Immunol 2006; 176:7062-7070.
23. Jelinek I, Laszlo V, Buzas E et al. Increased antigen presentation and Th1 polarization in genetically histamine-free mice. Int Immunol 2006; 19:51-58.
24. Koarai A, Ichinose M, Ishigaki-Suzuki S et al. Disruption of L-histidine decarboxylase reduces airway eosinophilia but not hyperresponsiveness. Am J Respir Crit Care Med 2003; 167:758-763.

25. Kozma GT, Losonczy G, Keszei M et al. Histamine deficiency in gene-targeted mice strongly reduces antigen-induced airway hyper-responsiveness, eosinophilia and allergen-specific IgE. Int Immunol 2003; 15:963-973.
26. Couillin I, Maillet I, Vargaftig BB et al. Arthropod-derived histamine-binding protein prevents murine allergic asthma. J Immunol 2004; 173:3281-3286.
27. Jutel M, Watanabe T, Klunker S et al. Histamine regulates T-cell and antibody responses by differential expression of H1 and H2 receptors. Nature 2001; 413:420-424.
28. Akdis CA, Simons FER. Histamine receptors are hot in immunopharmacology. Eur J Pharmacol 2006; 533:69-76.
29. Nakae S, Suto H, Berry GJ et al. Mast cell-derived TNF can promote Th17 cell-dependent neutrophil recruitment in ovalbumin-challenged OTII mice. Blood 2007; 109:3640-3648.
30. Gantner F, Sakai K, Tusche MW et al. Histamine H4 and H2 receptors control histamine—induced interleukin-16 release from human CD8$^+$ T-cells. J Pharmacol Exp Ther 2002; 303:300-307.
31. Sonobe Y, Nakane H, Watanabe T et al. Regulation of Con A-dependent cytokine production from CD4$^+$ and CD8$^+$ T-lymphocytes by autosecretion of histamine. Inflamm Res 2004; 53:87-92.
32. Kubo Y, Nakano K. Regulation of histamine synthesis in mouse CD4$^+$ and CD8$^+$ T-lymphocytes. Inflamm Res 1999; 48:149-153.
33. Radvany Z, Darvas Z, Kerekes K et al. H1 histamine receptor antagonist inhibits constitutive growth of Jurkat T-cells and antigen-specific proliferation of ovalbumin-specific murine T-cells. Semin Cancer Biol 2000; 10:41-45.
34. Dy M, Lebel B, Kamoun P et al. Histamine production during the anti-allograft response. Demonstration of a new lymphokine enhancing histamine synthesis. J Exp Med 1981; 153:293-309.
35. Schneider E, Lemoine F, Breton-Gorius J et al. IL-3-induced coexpression of IL-4, IL-6 and histidine decarboxylase mRNA in basophilic myelocytes enriched in sorted-Rh-bright bone marrow cells. Exp Haematol 1999; 27: 1010-1018.
36. Jacobs ER, Zeldin DC. The lung HEEs (and EETs) up. Am J Physiol Circ Physiol 2001; 280: H1-H10.
37. Schneider E, Ploemacher RE, Nabarra B et al. Mast cells and their committed precursors are not required for IL-3-induced histamine synthesis in murine bone marrow: characteristics of histamine-producing cells. Blood 1993; 81:1161-1169.
38. Dy M, Schneider E, Gastinel LN et al. Histamine-producing cell-stimulating activity. A biological activity shared by interleukin 3 and granulocyte-macrophage colony stimulating factor. Eur J Immunol 1987; 17:1243-1248.
39. Dy M, Lebel B. Skin allografts generate an enhanced production of histamine and HCSF by spleen cells in response to T-cell mitogens. J Immunol 1983; 130:2343-2347.
40. Kunzmann S, Mantel PY, Wohlfahrt JG et al. Histamine enhances TGF-ß1-mediated suppression of Th2 responses. FASEB J 2003; 17:1089-1095.
41. Morgan RK, McAllister B, Cross L et al. Histamine 4 receptor activation induces recruitment of FoxP3$^+$ T-cells and inhibits allergic asthma in a murine model. J Immunol 2007; 178:8081-8089.
42. Kronenberg M. Toward an understanding of NKT cell biology: progress and paradoxes. Ann Rev Immunol 2005; 23:877-900.
43. Bendelac A, Savage PB, Teyton L. The biology of NKT cells. Ann Rev Immunol 2007; 25:297-336.
44. Michel M-L, Leite-de-Moraes MC. Other sources of IL-17: iNKT cells. In: Quesniaux V, Ryffel B, Pavona F, eds. TH17 Cells: Role in inflammation and autoimmune disease. Series Progress in Inflammation Research. Basel: Birkhauser Publishing, 2009:39-48.
45. Leite de Moraes MC, Diem S, Michel ML et al. Histamine receptor H4 activation positively regulates in vivo IL-4 and IFN-γ production by invariant natural killer T-cells. J Immunol Cutting Edge 2009; 182:1233-1236.
46. Marone G, Granata F, Spadaro G et al. The histamine-cytokine network in allergic inflammation. J Allergy Clin Immunol 2003; 112:S83-S88.
47. Banu Y, Watanabe T. Augmentation of antigen receptor-mediated responses by histamine H1 receptor signaling. J Exp Med 1999; 189:673-682.
48. Elenkov IJ, Webster E, Papanicolaou DA et al. Histamine potently suppresses human IL-12 and stimulates IL-10 production via H2 receptors. J Immunol 1998; 161:2586-2593.
49. Tineke CTM, Van Der Pouw K, Snijders A et al. Histamine inhibits the production of interleukin-12 through interaction with H2 receptors. J Clin Invest 1998; 102:1866-1873.
50. Rocklin RE, Blidy A, Kamal M. Physicochemical characterization of human histamine induced suppressor factor. Cell Immunol 1983; 76:243-252.
51. Vannier E, Miller LC, Dinarello CA. Histamine suppresses gene expression and synthesis of tumor necrosis factor alpha via histamine H2 receptors. J Exp Med 1991; 174:281-284.

52. Wang KY, Arima N, Higuchi S et al. Switch of histamine receptor expression from H2 to H1 during differentiation of monocytes into macrophages. FEBS Letters 2000; 473:345-348.
53. Triggiani M, Petraroli A, Loffredo S et al. Differentiation of monocytes into macrophages induces the upregulation of histamine H1 receptor. J Allergy Clin Immunol 2007; 119:472-481.
54. Kimura S, Wang KY, Tanimoto A et al. Acute inflammatory reactions caused by histamine via monocytes/macrophages chronically participate in the initiation and progression of atherosclerosis. Pathol Int 2004; 54:465-474.
55. Sasaguri Y, Tanimoto A. Role of macrophage-derived histamine in atherosclerosis—chronic participation in the inflammatory response. J Atherosclerosis Thrombosis 2004; 11:122-130.
56. Dijkstra D, Leurs R, Chazot P et al. Histamine downregulates monocyte CCL2 production through the histamine H4 receptor. J Allergy Clin Immunol 2007; 179:300-307.
57. Takahashi HK, Morichika T, Iwagaki H et al. Histamine downregulates CD14 expression via H2 receptors on human monocytes. Clin Immunol 2003; 108:274-281.
58. Katoh N, Soga F, Nara T et al. Histamine induces the generation of monocyte-derived dendritic cells that express CD14 but not CD1α. J Invest Dermatol 2005; 125:753-760.
59. Itoh H, Takahashi HK, Iwagaki H et al. Effect of histamine on intercellular adhesion molecule-1 expression and production of interferon-gamma and interleukin-12 in mixed lymphocyte reaction stimulated with interleukin-18. Transplantation 2002; 74:864-870.
60. Soga F, Katoh N, Kishimoto S. Histamine prevents apoptosis in human monocytes. Clin Exp Allergy 2007; 37:323-330.
61. Laszlo V, Rothe G, Hegyesi H et al. Increased histidine decarboxylase expression during in vitro monocyte maturation; a possible role of endogenously synthesized histamine in monocyte/macrophage differentiation. Inflamm Res 2001; 50:428-434.
62. Tanaka S, Deai K, Inagaki M et al. Uptake of histamine by mouse peritoneal macrophages and a macrophage cell line, RAW264.7. Am J Physiol Cell Physiol 2003; 285:C592-C598.
63. Buhring HJ, Simmons PJ, Pudney M et al. The monoclonal antibody 97A6 defines a novel surface antigen expressed on human basophils and their multipotent and unipotent progenitors. Blood 1999; 94:2343-2356.
64. I L, Li Y, Reddel S et al. Identification of basophilic cells that express mast cell granule proteases in the peripheral blood of asthma, allergy and drug-reactive patients. J Immunol 1998; 161: 2439-2445.
65. Mitre E, Taylor RT, Kubofcik J et al. Parasite antigen-driven basophils are a major source of IL-4 in human filarial infections. J Immunol 2004; 172:2439-2445.
66. Kawakami T, Galli SJ. Regulation of mast cell and basophil function and survival by IgE. Nat Rev 2002; 2:773-786.
67. Hofstra CL, Desai PJ, Thurmond RL et al. Histamine H4 receptor mediates chemotaxis and calcium mobilization of mast cells. J Pharmacol Exp Ther 2003; 305: 1212-1221.
68. Lim HD, Van Rijn RM, Ling P et al. Evaluation of histamine H1-, H2- and H3-receptor ligands at the human histamine H4 receptor: identification of 4-methylhistamine as the first potent and selective H4 receptor agonist. J Pharmacol Exp Ther 2005; 314:1310-1321.
69. Godot V, Arock M, Garcia G et al. H4 histamine receptor mediates optimal migration of mast cell precursors to CXCL12. J Allergy Clin Immunol 2007; 120:827-834.
70. Schneider E, Machavoine F, Pleau JM et al. Organic Cation Transporter 3 modulates murine basophil functions by controlling intracellular histamine levels. J Exp Med 2005; 202:387-393.
71. Brandes LJ, Queen GM, Labella FS. Potent interaction of histamine and polyamines at microsomal cytochrome P450, nuclei and chromatin from rat hepatocytes. J Cell Biochem 1998; 69:233-243.
72. Labella FS, Brandes LJ. Interaction of histamine and other bioamines with cytochromes P450: implications for cell growth modulation and chemopotentiation by drugs. Semin Cancer Biol 2000; 10:47-53.
73. Buckland KF, Williams TJ, Conroy DM. Histamine induces cytoskeletal changes in human eosinophils via the H4 receptor. Br J Pharmacol 2003; 140:1117-1127.
74. O'Reilly M, Alpert R, Jenkinson S et al. Identification of a histamine H4 receptor on human eosinophils-role in eosinophil chemotaxis. J Recept Signal Transduct Res 2002; 22:431-448.
75. Ling P, Ngo K, Nguyen S et al. Histamine H4 receptor mediates eosinophil chemotaxis with cell shape change and adhesion molecule upregulation. Br J Pharmacol 2004; 142:161-171.
76. Nakayama T, Kato Y, Hieshima K et al. Liver-expressed chemokine/CC chemokine ligand 16 attracts eosinophils by interacting with histamine H4 receptor. J Immunol 2004; 173:2078-2083.
77. Hasala H, Janka-Junttila M, Moilanen E et al. Levocetirizine and cytokine production and apoptosis of human eosinophils. Allergy Asthma Proc 2007; 28:582-591.
78. Hasala H, Giembycz MA, Janka-Junttila M et al. Histamine reverses IL-5 afforded human eosinophil survival by inducing apoptosis: pharmacological evidence for a novel mechanism of action of histamine. Pulm Pharmacol Ther 2008; 21:222-233.

79. Sugata Y, Okano M, Fujiwara T et al. Histamine H4 receptor agonists have more activities than H4 agonism in antigen-specific human T-cell responses. Immunology 2007; 121:266-275.
80. Takeshita K, Bacon KB, Gantner F. Critical role of L-selectin and histamine H4 receptor in Zymosan-induced neutrophil recruitment from the bone marrow: comparison with carrageenan. J Pharmacol Exp Therapeutics 2004; 310:272-280.
81. Varga C, Horvarth K, Berko A et al. Inhibitory effects of histamine H4 receptor antagonists on experimental colitis in the rat. Eur J Pharmacol 2005; 522:30-138.
82. Tanaka S, Deai K, Konomi A et al. Expression of L-histidine decarboxylase in granules of elicited mouse polymorphonuclear leukocytes. Eur J Immunol 2004; 34: 1472-1482.
83. Xu X, Zhang D, Zhang D et al. Neutrophil histamine contributes to inflammation in mycoplasma pneumonia. J Exp Med 2006; 203:2907-2917.
84. Tuomisto L, Kilpelainen H, Riekkinen P. Histamine and histamine-N-methyltransferase in the CSF of patients with multiple sclerosis. Agents Actions 1983; 13:255-257.
85. Petersen IJ, Hansen U, Kristensen JK et al. Studies on mast cells and histamine release in psoriasis: the effect of ranitidine. Acta Derm Venereol 1998; 78:190-193.
86. Winterkamp S, Weidenhiller M, Otte P et al. Urinary excretion of N-methylhistamine as a marker of disease activity in inflammatory bowel disease. Am J Gastroenterol 2002; 97:3071-3077.
87. Raithel M, Matek M, Baenkler HW et al. Mucosal histamine content and histamine secretion in Crohn's disease, ulcerative colitis and allergic enteropathy. Int Arch Allergy Immunol 1995; 108:127-133.
88. Zhang M, Venable JD, Thurmond RL. The histamine H4 receptor in autoimmune disease. Expert Opin Investig Grugs 2006; 15:1443-1452.
89. Ibrahim MZM, Reder AT, Lawand R et al. The mast cells of the multiple sclerosis brain. J Neuroimmunol 1996; 70:131-138.
90. Toms R, Weiner HL, Johnson D. Identification of IgE-positive cells and mast cells in frozen sections of multiple sclerosis brains. J Neuroimmunol 1990; 30:169-177.
91. Zappulla JP, Arock M, Mars LT et al. Mast cells: new targets for multiple sclerosis therapy? J Neuroimmunol 2002; 131:5-20.
92. Kruger PG, Bo L, Myhr KM et al. Mast cells and multiple sclerosis: a light and electron microscopic study of mast cells in multiple sclerosis emphasizing staining procedures. Acta Neurol Scand 1990; 81:31-36.
93. Godfrey HP, Ilardi C, Engber W et al. Quantitation of human synovial mast cells in rheumatoid arthritis and other rheumatic diseases. Arthritis Rheum 1984; 27:852-856.
94. Gotis-Graham I, Smith MD, Parker A et al. Synovial mast cell responses during clinical improvement in early rheumatoid arthritis. Ann Rheum Dis 1998; 57:664-671.
95. Tetlow LC, Woolley DE. Distribution, activation and tryptase/chymase phenotype of mast cells in the rheumatoid lesion. Ann Rheum Dis 1995; 54:549-555.
96. Skopouli FN, Li L, Boumba D. Association of mast cells with fibrosis and fatty infiltration in the minor salivary glands of patient with Sjogren's syndrome. Clin Exp Rheumatol 1998; 16:63-65.
97. Sokol CL, Barton GM, Farr AG et al. A mechanism for the initiation of allergen-induced T helper type 2 responses. Nat Immunol 2008; 9:310-318.
98. Pedotti R, DeVoss JJ, Youssef S et al. Multiple elements of the allergic arm of the immune response modulate autoimmune demyelination. Proc Natl Acad Sci USA 2003; 100:1867-1872.
99. Ferber IA, Brocke S, Taylor-Edwards C et al. Mice with a disrupted IF-gamma gene are susceptible to the induction of experimental autoimmune encephalomyelitis (EAE). J Immunol 1996; 156:5-7.
100. Willenborg DO, Fordham S, Bernard CC et al. IFN-gamma plays a critical down-regulatory role in the induction and effector phase of myelin oligo-dendrocyte glycoprotein-induced autooimmune encephalomyelitis. J Immunol 1996; 157:3223-3227.
101. Cua DJ, Sherlock J, Chen Y et al. Interleukin-23 rather than interleukin-12 is the critical cytokine for autoimmune inflammation of the brain. Nature 2003; 421:744-748.
102. Teuscher C, Subramanian M, Noubade R et al. Central histamine H3 receptor signaling negatively regulates susceptibility to autoimmune inflammatory disease of the CNS. Proc Natl Acad Sci USA 2007; 104:10146-10151.
103. Ma RZ, Gao J, Meeker ND et al. Identification of Bphs, an autoimmune disease locus, as histamine receptor H1. Science 2002; 297:620-623.
104. Emerson MR, Orentas DM, Lynch SG et al. Activation of histamine H2 receptors ameliorates experimental allergic encephalomyelitis. Neuroreport 2002; 13:1407-1410.
105. Musio S, Gallo B, Scabeni S et al. A key regulatory role for histamine in experimental autoimmune encephalomyelitis: disease exacerbation in histidine decarboxylase-deficient mice. J Immunol 2006; 176:17-26.
106. Beghadadi W, Porcherie A, Schneider BS et al. Histamine H3 receptor-mediated signalling protects mice from cerebral malaria. Plos One 2009; 4:e6004

107. Noubade R, Milligan G, Zachary JF et al. Histamine receptor H1 is required for TCR-mediated p38 MAPK activation and optimal IFN-gamma production in mice. J Clin Invest 2007; 117:3507-3518.
108. Noubade R, Saligrama N, Spach K et al. Autoimmune disease-associated histamine receptor H1 alleles exhibit differential protein trafficking and cell surface expression. J Immunol 2008; 180:7471-7479.
109. Brodell LA, Beck LA, Saini SS. Pathophysiology of chronic urticaria. Ann Allergy Asthma Immunol 2008; 100:291-298.
110. De Swerdt A, Van Den Keybus C, Kasran A et al. Detection of basophil-activating IgG autoantibodies in chronic idiopathic urticaria by induction of CD63. J Allergy Clin Immunol 2005; 116:662-667.
111. Asero R, Riboldi P, Tedeschi A et al. Chronic urticaria: a disease at a crossroad between autoimmunity and coagulation. Autoimmunity Rev 2007; 7:71-76.
112. O'Donnell BF, Francis DM, Swana GT et al. Thyroid autoimmunity in chronic urticaria. Clin Lab Invest 2005; 153:331-335.
113. Chen WC, Chiang BL, Liu HE et al. Defective functions of circulating CD4+CD25+ and CD4+CD25– T-cells in patients with chronic ordinary urticaria. J Dermatol Sci 2008; 51:121-130.
114. Sun RS, Chen XH, Liu RQ et al. Autoantibodies to the high-affinity IgE receptor in patients and asthma. Asian Pac J Allergy Immunol 2008; 26:19-22.
115. Mąslinska D, Gujski M, Laure-Kamionowska M et al. Subcellular localization of histamine in articular cartilage chondrocytes of rheumatoid arthritis patients. Inflamm. Res 2004; 53S:S35-S36.
116. Buckley MG, Walters C, Wong WM et al. Mast cell activation in arthritis: detection of alpha- and beta-tryptase, histamine and eosinophil cationic protein in synovial fluid. Clin Sci (London) 1997; 93:363-370.
117. Malone DG, Irani AM, Schwartz LB et al. Mast cell numbers and histamine levels in synovial fluids from patients with diverse arthritides. Arthritis Rheum 1986; 29:956-963.
118. Woolley DE. The mast cell in inflammatory arthritis. N Engl J Med 2003; 348:1709-1711.
119. Tetlow LC, Woolley DE. Histamine stimulates the proliferation of human articular chondrocytes in vitro and is expressed by chondrocytes in osteoarthritic cartilage. Ann Rheum Dis 2003; 62:991-994.
120. Tetlow LC, Woolley DE. Histamine, histamine receptors (H1 and H2) and histidine decarboxylase expression by chondrocytes of osteoarthritic cartilage: an immunohistochemical study. Rheumatol Int 2005; 26:173-178.
121. Frewin DB, Cleland LG, Jonsson JR et al. Histamine levels in human synovial fluid. J Rheumatol 1986; 13:13-14.
122. Gujski M, Wojtecka-Lukasik E, Gajewski M et al. Is lymphocyte histamine involved in the pathogenesis of rheumatoid arthritis? Inflamm Res 2000; 49:S25-S26.
123. Ohki E, Suzuki M, Aoe T et al. Expression of histamine H4 receptor in synovial cells from rhumatoid arthritic patients. Biol Pharm Bull 2007; 30:2217-2220.
124. Adlesic M, Verdrengh M, Bokarewa M et al. Histamine in rheumatoid arthritis. Scand J Immunol 2007; 65:530-537.
125. Goren N, Leiros PC, Sterin-Borda L et al. Effect of histamine in autoimmune myocarditis mice. Int J Immunopharmacol 1994; 16:737-745.
126. Goren N, Sterin-Borda L, Leiros PC et al. Increases in cyclic AMP levels couple to H1 receptors in atria from autoimmune myocarditis mice. Cell Signal 1995; 7:759-764.
127. Goren N, Sterin-Borda L, Bartrons R et al. Detection of mRNA encoding H(1) receptor and INOS by RT-PCR in autoimmune myocarditis with special reference to changes in heart contractility. Int J Cardiol 2000; 76:165-172.
128. Higuchi H, Hara M, Yamamoto K et al. Mast cells play a critical role in the pathogenesis of viral myocarditis. Circulation 2008; 118:363-372.
129. Meiler F, Zumhehr J, Klunler S et al. In vivo swich to IL-10-secreting T regulatory cells in high dose allergen exposure. J Exp Med 2008; 205:2887-2898.

Chapter 10

Histamine in Neurotransmission and Brain Diseases

Saara Nuutinen and Pertti Panula*

Abstract

Apart from its central role in the mediation of allergic reactions, gastric acid secretion and inflammation in the periphery, histamine serves an important function as a neurotransmitter in the central nervous system. The histaminergic neurons originate from the tuberomamillary nucleus of the posterior hypothalamus and send projections to most parts of the brain. The central histamine system is involved in many brain functions such as arousal, control of pituitary hormone secretion, suppression of eating and cognitive functions. The effects of neuronal histamine are mediated via G-protein-coupled H_1-H_4 receptors. The prominent role of histamine as a wake-promoting substance has drawn interest to treat sleep-wake disorders, especially narcolepsy, via modulation of H_3 receptor function. Post mortem studies have revealed alterations in histaminergic system in neurological and psychiatric diseases. Brain histamine levels are decreased in Alzheimer's disease patients whereas abnormally high histamine concentrations are found in the brains of Parkinson's disease and schizophrenic patients. Low histamine levels are associated with convulsions and seizures. The release of histamine is altered in response to different types of brain injury: e.g. increased release of histamine in an ischemic brain trauma might have a role in the recovery from neuronal damage. Neuronal histamine is also involved in the pain perception. Drugs that increase brain and spinal histamine concentrations have antinociceptive properties. Histaminergic drugs, most importantly histamine H_3 receptors ligands, have shown efficacy in many animal models of the above-mentioned disorders. Ongoing clinical trials will reveal the efficacy and safety of these drugs in the treatment of human patients.

Histaminergic Neurons

The first findings of histamine in the brain date back to 1919 when John J. Abel isolated histamine from the pituitary.[1] However, histamine's role as a neurotransmitter became evident only several decades later when lesions of the lateral hypothalamic area were found to decrease the activity of histamine synthetizing enzyme, L-histidine decarboxylase (HDC).[2] Another decade went by before methods became available to directly demonstrate the localization of the histaminergic neurons in the brain.[3,4] Cell bodies of histaminergic neurons are localized in the tuberomamillary nucleus (TMN) of the posterior hypothalamus from where they send projections to essentially all areas of the central nervous system similar to other amines (Fig. 1).[5] The number of histamine-containing neurons is about 4000 in the rat[6] whereas in human brain histaminergic neurons are more numerous (>64,000).[7]

*Corresponding Author: Pertti Panula—Institute of Biomedicine, POB 63, 00014 University of Helsinki, Finland. Email: pertti.panula@helsinki.fi

Histamine in Inflammation, edited by Robin L. Thurmond.

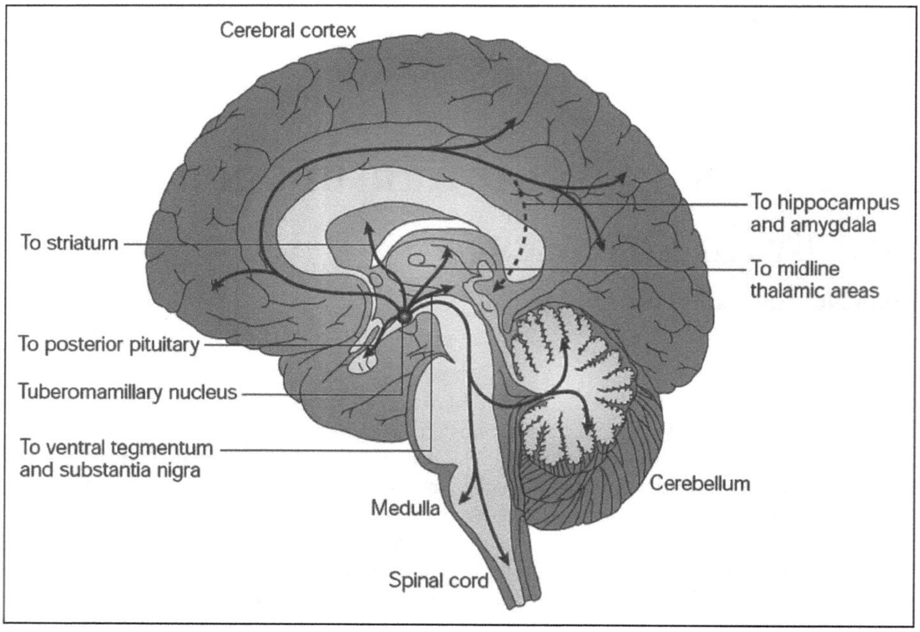

Figure 1. Histaminergic neurons and main projections in the human brain. Reprinted from: Haas H, Panula P. Nat Rev Neurosci 2003; 4:121-130.[5]

In addition to neurons, mast cells can produce histamine in the brain.[8] Indications of histamine synthesis in microglial cells also exists, but in vivo evidence of histamine produced by microglia is still missing.[9] It is noteworthy that brain ependymal cells also express HDC and can potentially synthesize histamine.[10] The role of this histamine is unknown, but it may be involved in regulation of stem cells located underneath the ependymal layer. Neural stem cells in vitro respond to both H_1R and H_2R receptor ligands.[11] Nonneuronal histamine has an important role in immune responses in the periphery and in the CNS and is covered elsewhere in this book. However, it should be noted that the source of brain histamine in some cases is difficult to detect. It is thus possible that both neuronal and nonneuronal histamine might regulate certain brain functions such as neuroinflammation.

Histamine Synthesis, Storage, Release and Catabolism

Histamine penetrates the brain poorly from blood, which protects the brain from many effects of blood-borne histamine. Histamine is synthetized from amino acid histidine by the specific histidine decarboxylase (HDC) enzyme in the brain.[2] The activity of HDC is highest in the hypothalamus where the histaminergic cell bodies are located, but HDC is also active in histaminergic nerve terminals.[3] The rate-limiting factor for histamine synthesis is the bioavailability of its precursor, histidine. Histamine is stored in vesicles in cell somata and especially in axon varicosities distinct from those containing GABA in the same cells.[12-14] Vesicular monoamine transporter 2 (VMAT-2) is responsible for the transport of histamine to the intracellular vesicles.[5] Upon arrival of action potentials histamine is released in a Ca^{2+}-dependent manner from the storage vesicles. In contrast to other amines, histaminergic synapses are rarely found in vertebrate nervous tissue and most histaminergic endings (varicosities) do not make close contact with postsynaptic sites.

Inactivation of histamine in the brain begins with a methylation reaction by histamine-N-methyltransferase.[15] Tele-methylhistamine undergoes oxidative deamination by monoamine oxidase B

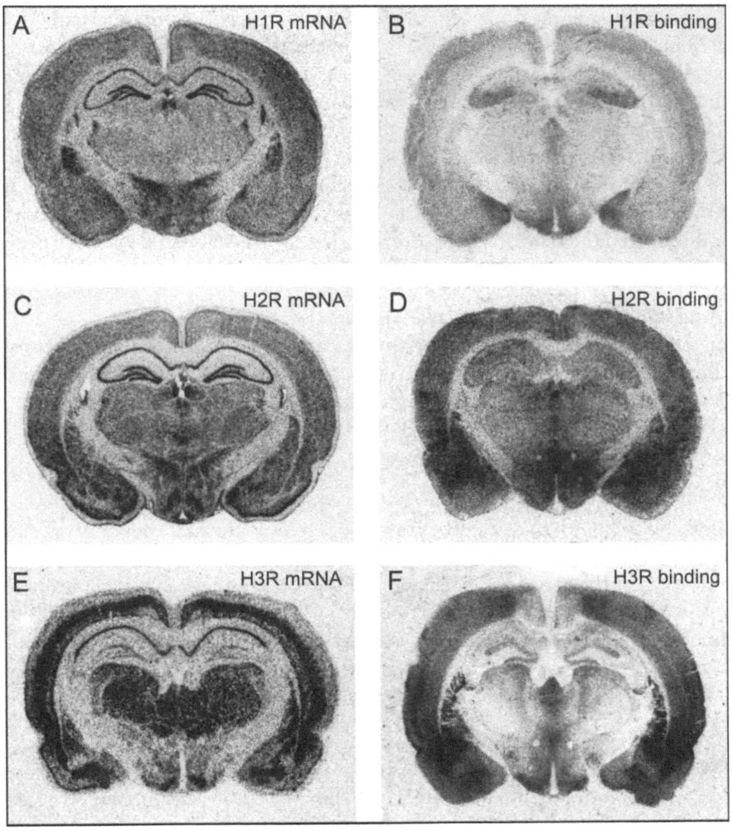

Figure 2. Distribution of histamine receptors in rat brain. Coronal sections at the level of the medial hypothalamus showing mRNA in situ hybridization and specific radioligand binding of H_1R (A, B), H_2R (C, D) and H_3R (E, F) receptors. Modified from: Haas H, Panula P. Nat Rev Neurosci 2003; 4:121-130.[5]

(MAO-B) into *t*-methyl-imidazolacetic acid.[16] Diamine oxidase (DAO) is the main histamine metabolizing enzyme in the peripheral tissues, but its activity in the brain is considerably low in basal conditions. However, it can be activated under conditions where methylation is inhibited.[17]

Histamine Receptors in the Brain

The actions of histamine in the brain are mediated through four G protein-coupled receptors, H_1R-H_4R (Fig. 2). H_1R and H_2R are postsynaptic and mediate mostly excitatory actions on neurons or potentiate excitatory inputs.[5] H_3R are located on histaminergic and other neurons on their somata, dendrites and axons.[18] On presynaptic sites H_3 autoreceptors inhibit the synthesis and release of histamine and H_3 heteroreceptors inhibit the release of other neurotransmitters.[19,20] The importance of postsynaptic H_3R on somata of neurons other than histaminergic neurons is not fully understood, but a recent study demonstrated that in striatum they are able to couple to dopamine D2 receptors and this interaction decreases the affinity of D2 receptors for its agonists.[21] An interesting property of H_3R is the high constitutive activity, which means spontaneous activity in the absence of histamine.[22] The constitutive activity has a potential regulatory role in the brain and several inverse agonists that are able to block this activity are currently in clinical trials to prove their efficacy in disorders such as Alzheimer's disease, schizophrenia, epilepsy, narcolepsy and obesity.[23,24] Another particular feature of

H_3R is the existence of multiple isoforms[25,26] derived from a single gene[27] by alternative splicing. The distribution of H_1R-H_3R is widespread in the brain as characterized by mRNA in situ hybridization and specific radioligand binding studies (Fig. 2). H_4R were first thought to be expressed only in the periphery, but recent studies suggest that H_4R is also expressed both in human and rat brain with highest levels of H_4R mRNA detected in the spinal cord.[28]

Histamine neurons are pacemakers that display a regular spontaneous firing with low frequency (1-4 Hz). When waking up, the firing of histaminergic neurons is increased.[5] During slow wave sleep the firing is low and no firing can be detected during rapid-eye-movement (REM) sleep. The inhibition of histaminergic neurons during sleep is mediated via GABAergic control from the ventrolateral preoptic area (VLPO).[29]

Physiological Role of Neuronal Histamine

Brain histamine is implicated in brain homeostasis and control of several neuroendocrine functions. Histamine has an important role in the control of behavioral state, biological rhythms, body weight, energy metabolism, thermoregulation, fluid balance, stress and reproduction.[5] In addition, histamine is implicated in higher brain functions such as sensory and motor functions, mood state, reward, learning and memory.

Arousal

Histamine has a prominent role in the control of arousal. The first cues of histamine as a waking substance came from the unwanted sedative side-effects of the first generation antihistamines that were able to cross the blood brain barrier. EEG recordings have shown that tuberomammillary neurons fire during wakefulness, but not during sleep.[30,31] In agreement, manipulation of the histaminergic system by H_3R antagonists to activate histaminergic neurons,[32] inhibition of histamine synthesis by alpha-fluoromethylhistidine[33] or histidine decarboxylase gene deletion (HDC knockout mice) leads to disturbances in sleep and waking state.[34]

Control of Pituitary Hormone Secretion

The role of histamine in the regulation of various endocrine functions is due to the effects of histamine on secretion of pituitary hormones.[35] This function is also in agreement with abundant expression of H_1R-H_3R in the hypothalamus (Fig. 2). Histamine regulates fluid balance via activation of H_1R localized on the neurons of supraoptic nucleus, which causes the release of vasopressin[36,37] and in turn induces antidiuresis.[38,39] Histaminergic neurons are also activated during parturition and lactation regulating the release of oxytocin and prolactin.[35,40,41] In addition, certain subgroups of histaminergic neurons are activated in response to stressful stimuli and control the release of adrenocorticotropic hormone.[42] Histamine also participates in the regulation of the release of growth hormone and thyrotropin-releasing hormone.[35]

Appetite and Body Weight

A large body of evidence links neuronal histamine to the regulation of appetite and body weight. First indications came from the appetite stimulating and weight increasing side-effects of first generation antipsychotics and antidepressants that had strong H_1R antagonist properties.[43] Later, several studies have demonstrated that histamine acts as an anorexigenic agent via stimulation of H_1R.[44] Histamine mediates the inhibitory effect of leptin on appetite via H_1R[45] confirmed by the complete absence of leptin-induced feeding suppression in H_1R knockout mice.[46,47] The effects of histamine on appetite are linked to various other neuroendocrine peptides such as orexins, neuropeptide Y, peptide YY and bombesin.[48] In addition to control of appetite, neuronal histamine affects metabolism by increasing lipolysis.[49,50] H_3R are promising targets to treat obesity since blockade of H_3R seems to be beneficial in decreasing energy intake, body weight and plasma triglycerides.[23] However, based on inconsistent results in H_3R antagonist studies, further investigations are needed to prove the potential of these drugs in the treatment of obesity and weight gain.

Histamine in Brain Diseases

Even though no pathological states have been selectively connected to deficits in the brain histamine system, alterations in histaminergic neurotransmission have been found in many neurological and psychiatric diseases such as sleep disorders, disorders of mood and cognition (schizophrenia, depression, Alzheimer's disease), movement disorders (Parkinson's disease), epilepsy, eating disorders, pain, neuroinflammation and addiction.[48] In the following parts of this chapter the role of histamine in some selected important brain diseases will be reviewed.

Sleep-Wake Disorders

Narcolepsy is a rare sleep disorder which is characterized by excessive daytime sleepiness that can be accompanied by manifestation of sudden loss of muscle tone triggered by emotional factors, referred to as cataplexy.[51] Pathophysiological studies have shown that narcolepsy is caused by the early loss of orexinergic neurons in the hypothalamus. Histaminergic neurons remain active during cataplexy whereas norepinephrine neurons stop firing and serotonin neurons lose much of their activity.[52] H_3R antagonists reduce sleepiness and cataplexy in animal models, probably due to the blockade of autoreceptors on histaminergic neurons resulting in increased release of histamine. Several compounds are being investigated in Phase II clinical trials for the treatment of narcolepsy.[23]

Due to the central role of histamine in the control of arousal and wake state, histamine receptors are potential targets for treatment of other types of sleep and wakefulness disorders as well. Doxepin is a tricyclic antidepressant that displays antagonistic effects on H_1R/H_2R in addition to the inhibitory action on norepinephrine and serotonin reuptake. The most common side-effect of doxepin is sedation and it has been shown to improve sleep quality in elderly patients suffering from insomnia.[53] Hypersomnia, on the other hand, could be treated by enhancing histaminergic activity. H_3R control histaminergic activity and histamine release and thus are promising targets to treat hypersomnia.

Alzheimer's Disease

Neuropathological studies have demonstrated deficits in the histaminergic system of Alzheimer's disease (AD) patients. Histamine and histidine decarboxylase levels are decreased in some key areas for cognition such as frontal cortex and hippocampus (Table 1).[54,55] Furthermore, numerous neurofibrillary tangles are found in the tuberomammillary nucleus of the AD brain and the number of large neurons in TMN is decreased, which may at least partly cause the histaminergic dysfunction in AD brain.[56,57] The number of H_1R ligand binding sites is decreased in the AD brain,[58] but interestingly H_3R levels seem to remain normal.[59] It is possible that decreased histaminergic activity may participate in the cognitive impairments of AD based on the ability of histamine to activate septohippocampal GABAergic neurons through both direct and indirect (cholinergic) mechanisms, which contribute to maintenance of hippocampal theta rhythm and thus cognition and memory.[60] A potentially important target for cognitive effects of H_3R ligands is also the thalamocortical system since the H_3R is expressed at particularly high levels in both the rat[61] and human[62] thalamus and cerebral cortex. Interestingly in the mouse, the H_3R expression is significantly lower, which may render studies with mice less relevant for modeling of human disease processes than those with rats.

Novel H_3R antagonists increase acetylcholine levels in cortical areas and hippocampus and improve performance in different cognition paradigms in experimental animals.[59,63-65] Ongoing clinical trials will show whether these compounds are effective in patients as well. In addition to memory-improving effects, H_3R antagonists show efficacy in attention and impulsivity, which makes them attractive candidates for the treatment of attention deficit hyperactivity disorder (ADHD) and cognitive deficits in schizophrenia (see below).

Table 1. Changes in histamine concentrations in different brain areas of Parkinson's disease and Alzheimer's disease patients

	Parkinson's Disease	Alzheimer's Disease
Caudate	**+28***	±0
Putamen	**+59***	−60
Substantia nigra pars compacta	**+101***	−27
Substantia nigra pars reticulata	+64	−23
Globus pallidus internum	**+134***	-
Globus pallidus externum	**+100***	-
Hypothalamus	+47	**−58***
Hippocampus	+16	**−57***
Frontal cortex	+17	−33
Temporal cortex	−5	**−47***
Occipital cortex	−18	−25

Data are expressed as a percentage change in histamine concentration from corresponding control brains. Asterisks refer to statistically significant differences from controls. Modified from Panula et al[54] and Rinne et al.[78]

Schizophrenia

Studies in humans and models in rodents support a role for histamine in the pathophysiology of schizophrenia. The levels of histamine's major metabolite, N-tele-methylhistamine, are elevated in the cerebrospinal fluid of schizophrenics[66] whereas the H_1R binding sites are decreased.[67,68] In agreement, repeated administration of methamphetamine which results in behavioral sensitization to dopamine agonists, a cardinal feature of schizophrenia, is accompanied by enhanced histamine release in rat brain.[69] Similar increases in histamine release are found in mice treated with phencyclidine.[70] Histamine H_3R radioligand binding is significantly increased postmortem in the prefrontal cortex of schizophrenics as compared to normal control, bipolar or depressive subjects.[71] Although this may be a consequence of drug treatment, lack of such differences in the temporal cortex of the same subjects suggests that increased H_3R radioligand binding may be directly related to the disease process. H_3R may play important roles in regulation of the thalamocortical system, which is essential for sensory systems and cognitive functions. H_3R mRNA is expressed at very high level in the dorsal thalamic nuclei of the human brain[72] and in layers IV and V of different subregions of the prefrontal cortex.[62]

Based on the alterations in the histaminergic system in schizophrenic patients, histaminergic drugs might be useful in the treatment of schizophrenia. Indeed, in preclinical studies, H_3R antagonists display antipsychotic effects by improving the deficits in sensomotoric gating in a prepulse inhibition of startle model and by reducing the hyperactivity induced by methamphetamine.[73] Some new H_3R antagonists also increase the release of dopamine in rat prefrontal cortex[59,63] which is regarded as beneficial since hypodopaminergic function in prefrontal cortex is associated with negative symptoms and cognitive deficits of schizophrenia.[74] Tiprolisant (BF2.649) is a novel H_3R antagonist that has an antipsychotic profile in animal models and shows efficacy in patients suffering from antipsychotic-induced weight gain. The effect of tiprolisant on cognitive functions is currently being investigated in clinical trials.[23] Interestingly several open studies have shown that the H_2R antagonist famotidine reduces negative symptoms in schizophrenics,[75,76] a finding that needs to be confirmed in controlled studies.

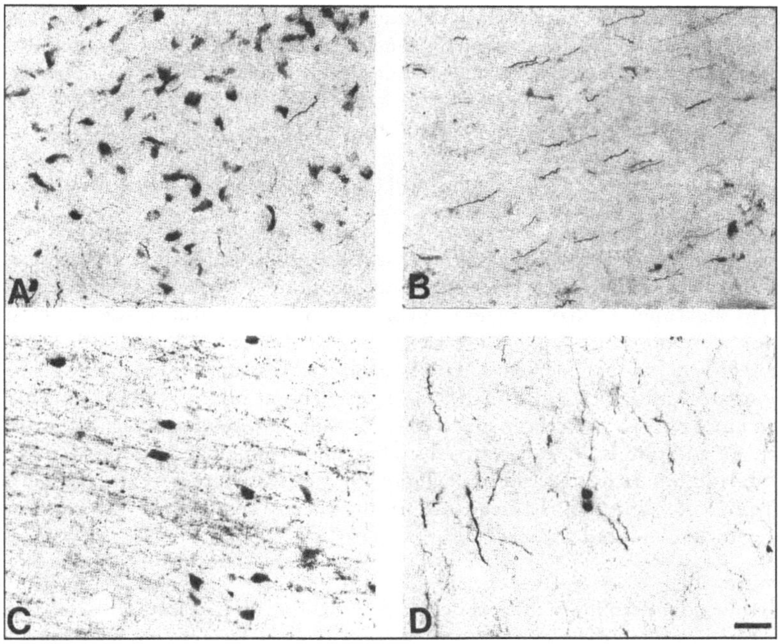

Figure 3. Histaminergic fibers in the substantia nigra of normal and Parkinson's disease brain. A,B) Histamine-immunoreactive fibers in normal brain substantia nigra pars compacta (A) and pars reticulata (B). C,D) Histaminergic fibers in Parkinson's disease brain substantia nigra pars compacta (C) and pars reticulata (D). Reproduced with permission from: Anichtchik OV et al. Exp Neurol 2000; 163:20-30,[77] ©2000 Elsevier.

Parkinson's Disease

Morphological changes and an increase in histaminergic fibers in the substantia nigra have been detected in Parkinson's disease (PD) patients (Fig. 3).[77] Also the histamine levels in the brains of Parkinson patients are increased in areas that regulate motor behavior such as the putamen, substantia nigra and globus pallidus (Table 1).[78] Polymorphism of Thr105Ile histamine-N-methyltransferase that leads to increased turnover of histamine is associated with PD risk.[79] Furthermore, H_3R ligand binding levels are increased[80] or unchanged[81] in PD patients. The H_3R binding is also increased in an animal model of PD where dopaminergic neurons are destroyed by a neurotoxic agent 6-OHDA.[82,83] H_3R regulate the release of GABA[84] and serotonin[85] in the direct and indirect striato-nigral movement pathways and might thus serve as possible drug targets in the treatment of Parkinson's disease and other movement disorders. H_3R antagonists improved motor coordination in 6-OHDA-lesioned rats.[86] In contrast, the H_3R agonist immepip, co-administered with L-dopa, decreased dyskinesia in nonhuman primate model of PD, whereas it increased parkinsonian disability when given alone.[87]

Since 1988[88] it has been known that inflammation and microglia may play important roles in PD. Both in animal models of PD and in PD patients, many characteristic features of neuroinflammation have been found.[89] An alteration of blood-brain barrier (BBB) function, characterized by increased permeability to both FITC-labeled albumin and horseradish peroxidase and neovascularization, follows an experimental lesion induced by intracerebral 6-OHDA.[90] Endothelial proliferation has also been found in human PD patients[91] and this seems to be associated with dysfunction of the BBB in PD patients as evidenced by uptake of (^{11}C)verapamil.[92] The alteration of BBB function may be a secondary phenomenon, since in experimental animals it is observed ipsilaterally to the 6-OHDA lesion.[90] Whether the alteration of BBB function is primary or secondary in PD, it is potentially important as dysfunctional BBB may allow entry of blood-borne toxins in the affected areas. Similarly, entrance of histidine might increase leading to larger than normal amounts of substrate available for HDC present in the nerve fibers. Histamine synthesis is largely regulated by substrate availability and unusually high histidine levels following, e.g., portacaval anastomosis are associated with very high brain histamine levels.[93] Alternatively, increased histamine in PD may be associated with neovascularization.

Epilepsy

Animal models and clinical observations have revealed that the brain histaminergic system has an inhibitory effect on seizures.[5] H_1R antihistamines show pro-convulsant effects particularly in children[94] and suppression of histaminergic activity promotes seizures in animal models.[95-98] In addition, development of epileptic behavior by kindling of the amygdala is increased in HDC- and H_1R-knockout mice.[99] Histamine levels in cerebrospinal fluid of children with febrile convulsions are significantly lower than those in the febrile children without convulsions.[100] Low histamine levels have also been detected in Krushinski-Molodkina rats that are prone to epilepsy.[101] In addition to its anti-convulsive effects, histamine has neuroprotective properties[102,103] and inhibits excitotoxic effects of glutamate.[103] H_3R antagonists have shown beneficial effects in different seizure models, such as electrically induced convulsions,[104,105] kindling[106] and pentylenetetrazole-induced seizures.[107]

Brain Injury

Involvement of histamine in the pathophysiology of brain injuries has been demonstrated in hypoxia,[108] trauma,[109,110] ischemia and stroke[111] or neoplasms.[112] For example, in rat brain, trauma leads to increased histamine levels both in the plasma and in the traumatized hemisphere.[110] In another study, fluid-percussion-evoked brain trauma caused changes in H_3R regulation demonstrated by receptor binding and mRNA in situ hybridization experiments.[109] In ischemic brain damage, histamine is released gradually over a long time frame.[113] Since preischemic administration of alpha-fluoromethylhistidine completely abolishes the ischemia-induced increase in the brain histamine, the source of histamine in cerebral ischemia is regarded to be neuronal.[113,114] An increased level of brain histamine may contribute to the amelioration of ischemic neuronal damage.[115] Thus, the emergence of neuronal injuries after ischemic events could be decreased by increasing brain histamine levels by histidine loading or H_3R blockade.

Pain

The descending histaminergic neurons originating from TMN project to areas related to pain perception[116] such as the dorsal raphe nucleus, periaqueductal gray region and dorsal horn of the spinal cord.[4,117,118] Several studies have demonstrated that histamine applied directly into CNS induces antinociception.[119-123] In line with this, brain histamine level reduction by alpha-fuoromethylhistidine or H_3R agonists enhances nociception.[123,124] Studies using histaminergic ligands[121] or mice lacking either H_1R or H_2R[125] suggest that H_1R and H_2R are responsible for the mediation of histamine's effects on pain perception.

Spinal H_3R have been also linked to antinociception based on studies with H_3R knockout mice and H_3R antagonists.[126,127] Interestingly, activation of H_3R seems to inhibit only certain types of pain such as pain induced by mechanical low-intensity stimulation.[128] The antinociceptive properties of novel H_4R antagonists[129] and recently described expression of H_4R in spinal cord[28] suggest a role for neuronal H_4R in pain perception as well.

Conclusion

The brain histaminergic system participates in the regulation of various brain functions, including sleep-wake cycle, energy and endocrine homeostasis and cognition. The effects of histamine in the brain are mediated via four histamine receptors (H_1R-H_4R), of which H_1R-H_3R are highly expressed. Changes in the neuronal histaminergic system are found in various brain disorders such as Alzheimer's disease, Parkinson's disease and schizophrenia, making histamine receptors promising targets for future drug therapies.

Acknowledgements

Original research in the authors' laboratory has been supported by the Academy of Finland.

References

1. Abel JJ, Kubota S. On the presence of histamine (-imidazolylethylamine) in the hypophysis cerebri and other tissues of the body and its occurence among the hydrolytic decomposition products of proteins. J Pharmacol Exp Ther 1919; 13:243-300.
2. Garbarg M, Barbin G, Bischoff S et al. Evidence for a specific decarboxylase involved in histamine synthesis in an ascending pathway in rat brain. Agents Actions 1974; 4:181.
3. Watanabe T et al. Evidence for the presence of a histaminergic neuron system in the rat brain: an immunohistochemical analysis. Neurosci Lett 1983; 39:249-254.
4. Panula P, Yang HY, Costa E. Histamine-containing neurons in the rat hypothalamus. Proc Natl Acad Sci. USA 1984; 81:2572-2576.
5. Haas H, Panula P. The role of histamine and the tuberomamillary nucleus in the nervous system. Nat Rev Neurosci 2003; 4:121-130.
6. Ericson H, Watanabe T, Kohler C. Morphological analysis of the tuberomammillary nucleus in the rat brain: delineation of subgroups with antibody against L-histidine decarboxylase as a marker. J Comp Neurol 1987; 263:1-24.
7. Panula P, Airaksinen MS, Pirvola U et al. A histamine-containing neuronal system in human brain. Neuroscience 1990; 34:127-132.
8. Martres MP, Baudry M, Schwartz JC. Histamine synthesis in the developing rat brain: evidence for a multiple compartmentation. Brain Res 1975; 83:261-275.
9. Katoh Y et al. Histamine production by cultured microglial cells of the mouse. Neurosci Lett 2001; 305:181-184.
10. Karlstedt K, Nissinen M, Michelsen KA et al. Multiple sites of L-histidine decarboxylase expression in mouse suggest novel developmental functions for histamine. Dev Dyn 2001; 221:81-91.
11. Molina-Hernandez A, Velasco I. Histamine induces neural stem cell proliferation and neuronal differentiation by activation of distinct histamine receptors. J Neurochem 2008; 106:706-717.
12. Kukko-Lukjanov TK, Panula P. Subcellular distribution of histamine, GABA and galanin in tuberomamillary neurons in vitro. J Chem Neuroanat 2003; 25:279-292.
13. Erickson JD, Schafer MK, Bonner TI et al. Distinct pharmacological properties and distribution in neurons and endocrine cells of two isoforms of the human vesicular monoamine transporter. Proc Natl Acad Sci USA 1996; 93:5166-5171.
14. Merickel A, Edwards RH. Transport of histamine by vesicular monoamine transporter-2. Neuropharmacology 1995; 34:1543-1547.
15. Reilly MA, Schayer RW. In vivo studies on histamine catabolism and its inhibition. Br J Pharmacol 1970; 38:478-489.
16. Hough LB, Domino EF. Tele-methylhistamine oxidation by type B monoamine oxidase. J Pharmacol Exp Ther 1979; 208:422-428.
17. Prell GD, Morrishow AM, Duoyon E et al. Inhibitors of histamine methylation in brain promote formation of imidazoleacetic acid, which interacts with GABA receptors. J Neurochem 1997; 68:142-151.
18. Arrang JM, Garbarg M, Schwartz JC. Auto-inhibition of brain histamine release mediated by a novel class (H3) of histamine receptor. Nature 1983; 302:832-837.
19. Schlicker E, Malinowska B, Kathmann M et al. Modulation of neurotransmitter release via histamine H3 heteroreceptors. Fundam Clin Pharmacol 1994; 8:128-137.
20. Brown RE, Stevens DR, Haas HL. The physiology of brain histamine. Prog Neurobiol 2001; 63:637-672.
21. Ferrada C et al. Interactions between histamine H3 and dopamine D2 receptors and the implications for striatal function. Neuropharmacology 2008; 55:190-197.
22. Morisset S et al. High constitutive activity of native H3 receptors regulates histamine neurons in brain. Nature 2000; 408:860-864.
23. Sander K, Kottke T, Stark H. Histamine H3 receptor antagonists go to clinics. Biol Pharm Bull 2008; 31:2163-2181.
24. Arrang JM, Morisset S, Gbahou F. Constitutive activity of the histamine H3 receptor. Trends Pharmacol Sci 2007; 28:350-357.
25. Drutel G et al. Identification of rat H3 receptor isoforms with different brain expression and signaling properties. Mol Pharmacol 2001; 59:1-8.

26. Rouleau A et al. Cloning and expression of the mouse histamine H3 receptor: evidence for multiple isoforms. J Neurochem 2004; 90:1331-1338.
27. Lovenberg TW et al. Cloning and functional expression of the human histamine H3 receptor. Mol Pharmacol 1999; 55:1101-1107.
28. Strakhova MI et al. Localization of histamine H(4) receptors in the central nervous system of human and rat. Brain Res 2008.
29. Sherin JE, Shiromani PJ, McCarley RW et al. Activation of ventrolateral preoptic neurons during sleep. Science 1996; 271:216-219.
30. Lin JS, Hou Y, Sakai K et al. Histaminergic descending inputs to the mesopontine tegmentum and their role in the control of cortical activation and wakefulness in the cat. J Neurosci 1996; 16:1523-1537.
31. Lin JS. Brain structures and mechanisms involved in the control of cortical activation and wakefulness, with emphasis on the posterior hypothalamus and histaminergic neurons. Sleep Med Rev 2000; 4:471-503.
32. Lin JS et al. Involvement of histaminergic neurons in arousal mechanisms demonstrated with H3-receptor ligands in the cat. Brain Res 1990; 523:325-330.
33. Kiyono S et al. Effects of alpha-fluoromethylhistidine on sleep-waking parameters in rats. Physiol Behav 1985; 34:615-617.
34. Parmentier R et al. Anatomical, physiological and pharmacological characteristics of histidine decarboxylase knock-out mice: evidence for the role of brain histamine in behavioral and sleep-wake control. J Neurosci 2002; 22:7695-7711.
35. Knigge U, Warberg J. The role of histamine in the neuroendocrine regulation of pituitary hormone secretion. Acta Endocrinol (Copenh) 1991; 124:609-619.
36. Armstrong WE, Sladek CD. Evidence for excitatory actions of histamine on supraoptic neurons in vitro: mediation by an H1-type receptor. Neuroscience 1985; 16:307-322.
37. Haas HL, Wolf P, Nussbaumer JC. Histamine: action on supraoptic and other hypothalamic neurones of the cat. Brain Res 1975; 88:166-170.
38. Bhargava KP, Kulshrestha VK, Santhakumari G et al. Mechanism of histamine-induced antidiuretic response. Br J Pharmacol 1973; 47:700-706.
39. Tuomisto L, Eriksson L, Fyhrquist F. Vasopressin release by histamine in the conscious goat. Eur J Pharmacol 1980; 63:15-24.
40. Libertun C, McCann SM. The possible role of histamine in the control of prolactin and gonadotropin release. Neuroendocrinology 1976; 20:110-120.
41. Hashimoto H, Noto T, Nakajima T. A study on the release mechanism of vasopressin and oxytocin. Neuropeptides 1988; 12:199-206.
42. Miklos IH, Kovacs KJ. Functional heterogeneity of the responses of histaminergic neuron subpopulations to various stress challenges. Eur J Neurosci 2003; 18:3069-3079.
43. Kalucy RS. Drug-induced weight gain. Drugs 1980; 19:268-278.
44. Jorgensen EA, Knigge U, Warberg J et al. Histamine and the regulation of body weight. Neuroendocrinology 2007; 86:210-214.
45. Morimoto T et al. Involvement of the histaminergic system in leptin-induced suppression of food intake. Physiol Behav 1999; 67:679-683.
46. Masaki T, Yoshimatsu H, Chiba S et al. Targeted disruption of histamine H1-receptor attenuates regulatory effects of leptin on feeding, adiposity and UCP family in mice. Diabetes 2001; 50:385-391.
47. Masaki T et al. Involvement of hypothalamic histamine H1 receptor in the regulation of feeding rhythm and obesity. Diabetes 2004; 53:2250-2260.
48. Haas HL, Sergeeva OA, Selbach O. Histamine in the nervous system. Physiol Rev 2008; 88:1183-1241.
49. Bugajski J, Janusz Z. Lipolytic responses induced by intracerebroventricular administration of histamine in the rat. Agents Actions 1981; 11:147-150.
50. Yoshimatsu H et al. Histidine induces lipolysis through sympathetic nerve in white adipose tissue. Eur J Clin Invest 2002; 32:236-241.
51. Dauvilliers Y, Arnulf I, Mignot E. Narcolepsy with cataplexy. Lancet 2007; 369:499-511.
52. John J, Wu MF, Boehmer LN et al. Cataplexy-active neurons in the hypothalamus: implications for the role of histamine in sleep and waking behavior. Neuron 2004; 42:619-634.
53. Scharf M et al. Efficacy and Safety of Doxepin 1 mg, 3 mg and 6 mg in Elderly Patients With Primary Insomnia: a Randomized, Double-Blind, Placebo-Controlled Crossover Study. J Clin Psychiatry 2008.
54. Panula P et al. Neuronal histamine deficit in Alzheimer's disease. Neuroscience 1998; 82:993-997.
55. Mazurkiewicz-Kwilecki IM, Nsonwah S. Changes in the regional brain histamine and histidine levels in postmortem brains of Alzheimer patients. Can J Physiol Pharmacol 1989; 67:75-78.
56. Nakamura S et al. Loss of large neurons and occurrence of neurofibrillary tangles in the tuberomammillary nucleus of patients with Alzheimer's disease. Neurosci Lett 1993; 151:196-199.
57. Airaksinen MS, Reinikainen K, Riekkinen P et al. Neurofibrillary tangles and histamine-containing neurons in Alzheimer hypothalamus. Agents Actions 1991; 33:104-107.

58. Higuchi M et al. Histamine H(1) receptors in patients with Alzheimer's disease assessed by positron emission tomography. Neuroscience 2000; 99:721-729.
59. Medhurst AD et al. GSK189254, a novel H3 receptor antagonist that binds to histamine H3 receptors in Alzheimer's disease brain and improves cognitive performance in preclinical models. J Pharmacol Exp Ther 2007; 321:1032-1045.
60. Xu C et al. Histamine innervation and activation of septohippocampal GABAergic neurones: involvement of local ACh release. J Physiol 2004; 561:657-670.
61. Jin C, Lintunen M, Panula P. Histamine H(1) and H(3) receptors in the rat thalamus and their modulation after systemic kainic acid administration. Exp Neurol 2005; 194:43-56.
62. Jin CY, Panula P. The laminar histamine receptor system in human prefrontal cortex suggests multiple levels of histaminergic regulation. Neuroscience 2005; 132:137-149.
63. Ligneau X et al. BF2.649 (1-{3-(3-(4-Chlorophenyl)propoxy)propyl}piperidine, hydrochloride), a nonimidazole inverse agonist/antagonist at the human histamine H3 receptor: preclinical pharmacology. J Pharmacol Exp Ther 2007; 320:365-375.
64. Fox GB et al. Pharmacological properties of ABT-239 (4-(2-{2-((2R)-2-Methylpyrrolidinyl)ethyl}-benzofuran-5-yl)benzonitrile): II. Neurophysiological characterization and broad preclinical efficacy in cognition and schizophrenia of a potent and selective histamine H3 receptor antagonist. J Pharmacol Exp Ther 2005; 313:176-190.
65. Cowart M et al. 4-(2-(2-(2(R)-methylpyrrolidin-1-yl)ethyl)benzofuran-5-yl)benzonitrile and related 2-aminoethylbenzofuran H3 receptor antagonists potently enhance cognition and attention. J Med Chem 2005; 48:38-55.
66. Prell GD et al. Histamine metabolites in cerebrospinal fluid of patients with chronic schizophrenia: their relationships to levels of other aminergic transmitters and ratings of symptoms. Schizophr Res 1995; 14:93-104.
67. Nakai T et al. Decreased histamine H1 receptors in the frontal cortex of brains from patients with chronic schizophrenia. Biol Psychiatry 1991; 30:349-356.
68. Iwabuchi K et al. Histamine H1 receptors in schizophrenic patients measured by positron emission tomography. Eur Neuropsychopharmacol 2005; 15:185-191.
69. Morisset S et al. Acute and chronic effects of methamphetamine on tele-methylhistamine levels in mouse brain: selective involvement of the D(2) and not D(3) receptor. J Pharmacol Exp Ther 2002; 300:621-628.
70. Itoh Y, Oishi R, Nishibori M et al. Phencyclidine and the dynamics of mouse brain histamine. J Pharmacol Exp Ther 1985; 235:788-792.
71. Jin C, Anichtchik O, Panula P. Altered histamine H3 receptor radioligand binding in postmortem brain samples from subjects with psychiatric diseases. Br J Pharmacol, In press 2009.
72. Jin CY, Kalimo H, Panula P. The histaminergic system in human thalamus: correlation of innervation to receptor expression. Eur J Neurosci 2002; 15:1125-1138.
73. Browman KE et al. Enhancement of prepulse inhibition of startle in mice by the H3 receptor antagonists thioperamide and ciproxifan. Behav Brain Res 2004; 153:69-76.
74. Esbenshade TA et al. The histamine H3 receptor: an attractive target for the treatment of cognitive disorders. Br J Pharmacol 2008; 154:1166-1181.
75. Kaminsky R, Moriarty TM, Bodine J et al. Effect of famotidine on deficit symptoms of schizophrenia. Lancet 1990; 335:1351-1352.
76. Martinez MC. Famotidine in the management of schizophrenia. Ann Pharmacother 1999; 33:742-747.
77. Anichtchik OV, Rinne JO, Kalimo H et al. An altered histaminergic innervation of the substantia nigra in Parkinson's disease. Exp Neurol 2000; 163:20-30.
78. Rinne JO et al. Increased brain histamine levels in Parkinson's disease but not in multiple system atrophy. J Neurochem 2002; 81:954-960.
79. Agundez JA et al. Nonsynonymous polymorphisms of histamine-metabolising enzymes in patients with Parkinson's disease. Neuromolecular Med 2008; 10:10-16.
80. Anichtchik OV, Peitsaro N, Rinne JO et al. Distribution and modulation of histamine H(3) receptors in basal ganglia and frontal cortex of healthy controls and patients with Parkinson's disease. Neurobiol Dis 2001; 8:707-716.
81. Goodchild RE et al. Distribution of histamine H3-receptor binding in the normal human basal ganglia: comparison with Huntington's and Parkinson's disease cases. Eur J Neurosci 1999; 11:449-456.
82. Ryu JH, Yanai K, Watanabe T. Marked increase in histamine H3 receptors in the striatum and substantia nigra after 6-hydroxydopamine-induced denervation of dopaminergic neurons: an autoradiographic study. Neurosci Lett 1994; 178:19-22.
83. Anichtchik OV et al. Modulation of histamine H3 receptors in the brain of 6-hydroxydopamine-lesioned rats. Eur J Neurosci 2000; 12:3823-3832.

84. Garcia M, Floran B, Arias-Montano JA et al. Histamine H3 receptor activation selectively inhibits dopamine D1 receptor-dependent (3H)GABA release from depolarization-stimulated slices of rat substantia nigra pars reticulata. Neuroscience 1997; 80:241-249.
85. Threlfell S et al. Histamine H3 receptors inhibit serotonin release in substantia nigra pars reticulata. J Neurosci 2004; 24:8704-8710.
86. Nowak P et al. Histamine H(3) receptor ligands modulate L-dopa-evoked behavioral responses and L-dopa derived extracellular dopamine in dopamine-denervated rat striatum. Neurotox Res 2008; 13:231-240.
87. Gomez-Ramirez J, Johnston TH, Visanji NP et al. Histamine H3 receptor agonists reduce L-dopa-induced chorea, but not dystonia, in the MPTP-lesioned nonhuman primate model of Parkinson's disease. Mov Disord 2006; 21:839-846.
88. McGeer PL, Itagaki S, Boyes BE et al. Reactive microglia are positive for HLA-DR in the substantia nigra of Parkinson's and Alzheimer's disease brains. Neurology 1988; 38:1285-1291.
89. Whitton PS. Inflammation as a causative factor in the aetiology of Parkinson's disease. Br J Pharmacol 2007; 150:963-976.
90. Carvey PM et al. 6-Hydroxydopamine-induced alterations in blood-brain barrier permeability. Eur J Neurosci 2005; 22:1158-1168.
91. Faucheux BA, Bonnet AM, Agid Y et al. Blood vessels change in the mesencephalon of patients with Parkinson's disease. Lancet 1999; 353:981-982.
92. Kortekaas R et al. Blood-brain barrier dysfunction in parkinsonian midbrain in vivo. Ann Neurol 2005; 57:176-179.
93. Fogel WA et al. Neuronal storage of histamine in the brain and tele-methylimidazoleacetic acid excretion in portocaval shunted rats. J Neurochem 2002; 80:375-382.
94. Haruyama W et al. The relationship between drug treatment and the clinical characteristics of febrile seizures. World J Pediatr 2008; 4:202-205.
95. Yokoyama H et al. Histamine levels and clonic convulsions of electrically-induced seizure in mice: the effects of alpha-fluoromethylhistidine and metoprine. Naunyn Schmiedebergs Arch Pharmacol 1992; 346:40-45.
96. Chen Z, Li WD, Zhu LJ et al. Effects of histidine, a precursor of histamine, on pentylenetetrazole-induced seizures in rats. Acta Pharmacol Sin 2002; 23:361-366.
97. Fujii Y, Tanaka T, Harada C et al. Epileptogenic activity induced by histamine H(1) antagonists in amygdala-kindled rats. Brain Res 2003; 991:258-261.
98. Yokoyama H, Sato M, Iinuma K et al. Centrally acting histamine H1 antagonists promote the development of amygdala kindling in rats. Neurosci Lett 1996; 217:194-196.
99. Hirai T et al. Development of amygdaloid kindling in histidine decarboxylase-deficient and histamine H1 receptor-deficient mice. Epilepsia 2004; 45:309-313.
100. Kiviranta T, Tuomisto L, Airaksinen EM. Histamine in cerebrospinal fluid of children with febrile convulsions. Epilepsia 1995; 36:276-280.
101. Onodera K, Tuomisto L, Tacke U et al. Strain differences in regional brain histamine levels between genetically epilepsy-prone and resistant rats. Methods Find Exp Clin Pharmacol 1992; 14:13-16.
102. Kukko-Lukjanov TK et al. Histaminergic neurons protect the developing hippocampus from kainic acid-induced neuronal damage in an organotypic coculture system. J Neurosci 2006; 26:1088-1097.
103. Dai H et al. Histamine protects against NMDA-induced necrosis in cultured cortical neurons through H receptor/cyclic AMP/protein kinase A and H receptor/GABA release pathways. J Neurochem 2006; 96:1390-1400.
104. Yokoyama H, Onodera K, Iinuma K et al. Effect of thioperamide, a histamine H3 receptor antagonist, on electrically induced convulsions in mice. Eur J Pharmacol 1993; 234:129-133.
105. Yokoyama H et al. Clobenpropit (VUF-9153), a new histamine H3 receptor antagonist, inhibits electrically induced convulsions in mice. Eur J Pharmacol 1994; 260:23-28.
106. Harada C et al. Intracerebroventricular administration of histamine H3 receptor antagonists decreases seizures in rat models of epilepsia. Methods Find Exp Clin Pharmacol 2004; 26:263-270.
107. Vohora D, Pal SN, Pillai KK. Histamine and selective H3-receptor ligands: a possible role in the mechanism and management of epilepsy. Pharmacol Biochem Behav 2001; 68:735-741.
108. Dux E et al. The blood-brain barrier in hypoxia: ultrastructural aspects and adenylate cyclase activity of brain capillaries. Neuroscience 1984; 12:951-958.
109. Lozada A, Maegele M, Stark H et al. Traumatic brain injury results in mast cell increase and changes in regulation of central histamine receptors. Neuropathol Appl Neurobiol 2005; 31:150-162.
110. Mohanty S et al. Role of histamine in traumatic brain edema. An experimental study in the rat. J Neurol Sci 1989; 90:87-97.
111. Waskiewicz J, Molchanova L, Walajtys-Rode E et al. Hypoxia and ischemia modifies histamine metabolism and transport in brain synaptosomes. Resuscitation 1988; 16:287-293.

112. Lefranc F, YeatonP, Brotchi J et al. Cimetidine, an unexpected anti-tumor agent and its potential for the treatment of glioblastoma (review). Int J Oncol 2006; 28:1021-1030.
113. Adachi N, Itoh Y, Oishi R et al. Direct evidence for increased continuous histamine release in the striatum of conscious freely moving rats produced by middle cerebral artery occlusion. J Cereb Blood Flow Metab 1992; 12:477-483.
114. Adachi N, Oishi R, Saeki K. Changes in the metabolism of histamine and monoamines after occlusion of the middle cerebral artery in rats. J Neurochem 1991; 57:61-66.
115. Adachi N. Cerebral ischemia and brain histamine. Brain Res Brain Res Rev 2005; 50:275-286.
116. Basbaum AI, Fields HL. Endogenous pain control systems: brainstem spinal pathways and endorphin circuitry. Annu Rev Neurosci 1984; 7:309-338.
117. Panula P et al. Histamine-immunoreactive nerve fibers in the mammalian spinal cord. Brain Res 1989; 484:234-239.
118. Panula P, Pirvola U, Auvinen S et al. Histamine-immunoreactive nerve fibers in the rat brain. Neuroscience 1989; 28:585-610.
119. Parolaro D et al. Histamine as a central modulator of rat intestinal transit. J Pharmacol Exp Ther 1989; 249:324-328.
120. Bhattacharya SK, Parmar SS. Antinociceptive effect of intracerebroventricularly administered histamine in rats. Res Commun Chem Pathol Pharmacol 1985; 49:125-136.
121. Thoburn KK, Hough LB, Nalwalk JW et al. Histamine-induced modulation of nociceptive responses. Pain 1994; 58:29-37.
122. Glick SD, Crane LA. Opiate-like and abstinence-like effects of intracerebral histamine administration in rats. Nature 1978; 273:547-549.
123. Malmberg-Aiello P, Lamberti C, Ghelardini C et al. Role of histamine in rodent antinociception. Br J Pharmacol 1994; 111:1269-1279.
124. Malmberg-Aiello P, Lamberti C, Ipponi A et al. Evidence for hypernociception induction following histamine H1 receptor activation in rodents. Life Sci 1998; 63:463-476.
125. Mobarakeh JI et al. Enhanced antinociception by intracerebroventricularly administered orexin A in histamine H1 or H2 receptor gene knockout mice. Pain 2005; 118:254-262.
126. Cannon KE, Leurs R, Hough LB. Activation of peripheral and spinal histamine H3 receptors inhibits formalin-induced inflammation and nociception, respectively. Pharmacol Biochem Behav 2007; 88:122-129.
127. Cannon KE et al. Activation of spinal histamine H3 receptors inhibits mechanical nociception. Eur J Pharmacol 2003; 470:139-147.
128. Cannon KE, Hough LB. Inhibition of chemical and low-intensity mechanical nociception by activation of histamine H3 receptors. J Pain 2005; 6:193-200.
129. Coruzzi G, Adami M, Guaita E et al. Anti-inflammatory and antinociceptive effects of the selective histamine H4-receptor antagonists JNJ7777120 and VUF6002 in a rat model of carrageenan-induced acute inflammation. Eur J Pharmacol 2007; 563:240-244.

Chapter 11

Histamine in Normal and Malignant Cell Proliferation

Andras Falus,* Zoltán Pós and Zsuzsanna Darvas

Abstract

Histamine is a biogenic amine widely distributed throughout the body. Given the observations that histamine can be induced and made available in an unstored diffusible form in tissues undergoing rapid growth (such as tumors and regenerating liver), it could have a role beyond inflammatory and allergic responses.

Histamine and Cell Proliferation

It is an old clinical/pathological observation that the risk of the development of cancerous processes is increased in tissue adjacent to areas of long-standing inflammation. Intuitively researchers and clinicians have long ago deduced that abnormal inflammatory responses and in particular those that are persistent, may promote the development of certain tumors. Recently, researchers have generated a more detailed picture about the molecular background of this empirical observation. It has been demonstrated that inflammatory mediators, which under normal circumstances are activated only for a short time and then are rapidly down-regulated, can, when improperly regulated, support the development, invasivity or angiogenic activity of some tumors. Convincing evidence is available showing that if production of inflammatory mediators is prolonged or their normal functions become modified, they can provide support for both tumor formation and progression. In accordance with these observations an attempt to redefine the paradigm of solid tumors has taken place, with new the interpretation of tumors as "wounds that never heal"—referring to an abnormally long and drawn-out regeneration-like process that is frequently present around progressing tumors. Histamine may have a role in this process since overexpression of histidine decarboxylase (HDC) has been detected in a wide range of tumors, including colon, breast, stomach, lung cancer and leukemia. On the other hand neoangiogenesis and antitumor immune responses are also clearly influenced by histamine and can modify local tumor growth. In this chapter we attempt to summarize the most well known effects of histamine on normal and tumor cell growth.

Histamine in Normal Cell Proliferation

The hypothesis that histamine could be involved in cell differentiation and proliferation was proposed in the 1960s and still remains controversial. In this section the discussion will be confined to "normal proliferation", that is proliferation where the normal highly sophisticated systems for controlling proliferation are functionally intact. Situations where proliferation control is corrupted, damaged or even completely abolished will be dealt with in the section on malignant proliferation.

*Corresponding Author: Andras Falus—Department of Genetics, Cell- and Immunobiology, Semmelweis University, H-1089, Budapest, Nagyvárad ter 4, Hungary.
Email: faland@dgci.sote.hu

Histamine in Inflammation, edited by Robin L. Thurmond.

In the following section an attempt will be made to overview the large amount of available data regarding the effects of histamine on cell proliferation. It is a widely accepted fact that proliferation—being one of the most common phenomena in living organisms—represents the very fundament of a great number of vital processes such as reproduction, growth, differentiation, functional and structural repair, regeneration, hematopoiesis and the immune response. Furthermore, as a consequence of its vast impact on virtually all physiological processes, proliferation stands under the strict control of probably the most secured regulative systems that evolution has equipped the species with. Obviously, proliferation cannot be interpreted within narrow borders as in most cases it means far more than simply cell reproduction. Proliferation often represents only an initiation step and is part of more complicated processes such as wound healing and immune responses.

Gastric Mucosa

Histamine is involved in the regulation of diverse physiological and pathological processes in the gastric mucosa. It is generally accepted that histamine plays a central role in mediating gastrin-stimulated gastric acid secretion of the stomach. Gastrin acts indirectly, as it induces HDC-expression, enhances HDC activity and forces histamine release from ECL (enterochromaffin-like) cells in the mucosa that in turn stimulates acid secretion in an H_2R dependent manner. In addition, gastrin is able to induce massive proliferation of several cell types within the mucosa. It has been proposed that gastrin-induced mucosal proliferation could in fact be a histamine/ECL cell-elicited response as well. The first known targets of the proliferation-supporting effects of gastrin were the mucus secreting cells of the mucosa. In a recent study, however, somewhat unexpected results were reported regarding the induction of gastrin-mediated mucosal proliferation.[2] In this work the impact of treatments with histamine H_1 receptor (H_1R), histamine H_2 receptor (H_2R) and histamine H_3 receptor (H_3R) antagonists (mepyramine, famotidine and clobenpropit, respectively) was investigated. In consensus with an earlier study[1] it was reported, that although signals coming through H_2R or H_1R are probably irrelevant, inhibition of H_3R expressed mostly on neurons blocked proliferation and emigration of stem cells of the fundic mucosa. In complete agreement with this, experimental H_3R activation via its specific agonist, (R)-alpha-methylhistamine, resulted in increased number, volume and secretory activity of differentiated mucus-secreting cells on the mucosal surface.[2]

In addition to mucus-secreting cells it is possible that gastrin-induced ECL cells could be targets of their own histamine secretion as well. The observation that ECL cells—being highly sensitive to gastrin induced proliferation—express considerable amounts of H_1R and H_2R support this possibility and it was suggested that H_1R acts as a proliferation promoting-factor in emerging ECLomas. Finally, some light was shed in this area owing to the work of two independent groups studying H_2R antagonist-mediated suppression of acid secretion.[3,4] Sustained blockade of acid secretion increases stomach pH, which evokes gastrin production as compensation. However, the prolongation of this hypergastrinaemia aberrantly increases tissue histamine content and overall mucosal thickness due to gastric cell and ECL-cell hyperplasia. The two groups independently concluded that although H_1R-coupled signals are unable to influence gastrin levels, gastric cell proliferation and, under normal physiological circumstances, ECL cell proliferation, this situation is changed during sustained suppression of acid secretion. In such cases, H_1R mediated signals support the proliferation of ECL cells.[3,4]

The third possible question is the putative connection between histamine and regenerative proliferation during the removal of mucosal damage associated with massive stress or ulcers. Ulcer healing can be accelerated by careful H_2R antagonist treatment owing to its moderating impact on acid secretion. Interestingly, in an in vitro experimental setting where the deleterious effect of gastric acid secretion was excluded, H_2R antagonist treatment remained effective by enhancing regeneration-like processes within a stimulated ulcer. It was observed that ulcer-like "wounds" scratched into confluent monolayers of MKN 28 gastric adenocarcinoma cells "healed" faster in H_2R-antagonist treated cultures than in untreated control ones. Accelerated healing was a result of a faster proliferation of cells neighboring the damaged areas, rather than enhanced migration to the "wounds".[5] Whether H_2R blockade could foster ulcer healing in this way in vivo remains to be elucidated. The already mentioned data suggesting that H_2R antagonists in healthy animals fail to influence mucosal proliferation may be insufficient to clearly rule out this possibility. As we have seen for examples with H_1R and ECL cells, the proliferation-inducing efficiency of histamine through the same receptor can be very different under different physiological situations. While H_1R remained irrelevant regarding proliferation of ECL-cells under normal physiological circumstances, they gained relevance in acid suppression. It is possible that similar differences can be observed with the roles played by H_2R in a healthy stomach versus an ulcer-bearing one.

Hematopoietic Proliferation

After initial work in the late 1970s showing that histamine was able to induce hematopoietic stem cell proliferation by promoting resting bone marrow cell progression from the G0 to the S phase via H_2R,[6] a real rush broke out in searching for further effects of histamine on hematopoiesis. At first it was suggested that histamine-mediated signals through H_2R may induce proliferation of certain precursor cell types within the myeloid lineage. At low concentrations (10^{-8} M) histamine was able to enhance granulocyte precursor proliferation and colony-forming unit activity (CFU-C) resulting in an increased number of differentiated neutrophils. The effect described, however, was limited to granulocytes, as the monocyte-macrophage lineage remained unaffected.[7]

The possible existence of such a differentiated effect of histamine on distinct hematopoietic precursors was again suggested by a report investigating LPS-provoked induction of HDC-expression in several tissues. It was found that G-CSF and GM-CSF could specifically evoke rapid HDC-activation in the bone marrow in contrast to other cytokines strongly coupled with inflammatory responses and characterized by broad HDC-activating properties like TNF-α, TNF-β, IL-1α and IL-1β. However, M-CSF does not transmit such signals, emphasizing the proposed differences between the sensitivity of granulocyte and macrophage precursor proliferation to histamine.[8] Later Dy and coworkers suggested that instead of exogenous histamine signals it might be advisable to consider nascent (i.e., newly synthesized) histamine from the hematopoietic bone marrow cell itself.[9] After stimulation with IL-3, a cytokine known to promote hemotopoiesis, bone marrow derived colony-forming units in the spleen (CFU-S) strongly up-regulated histamine production and this was apparently coupled with their entry in the cell cycle. This remarkable temporal association was proven to be functional as well, as inhibition of histamine synthesis with α-fluoromethylhistidine (αFMH) or blocking histamine signaling via H_2R strongly impeded IL-3 induced proliferation.[9] Further research by this group provided evidence for a similar requirement of nascent histamine synthesis for hematopoietic proliferation initiated by a combined treatment with IL-1 and GM-CSF.[10] Interestingly, neither IL-1 nor GM-CSF could evoke proliferation alone, although both supported survival of CFU-S cells under some circumstances. Regarding GM-CSF, this result was far more surprising: how is it possible to stimulate histamine synthesis in bone marrow cells on the one hand (as noted above), but not to induce proliferation on the other?

Although the exact background of these phenomena was not fully clarified, a theory was proposed that GM-CSF acts as a factor modulating not only histamine synthesis, but also diminishes the H_2R/H_1R ratio on the surface of the precursor cells. This reduces the proliferation-promoting capacity of histamine since H_2R and H_1R have opposite influences on the proliferation of stem cells. This proliferation-blocking regulatory effect of GM-CSF on histamine receptors was suggested to be alleviated by simultaneous IL-1 signals thereby rendering the cells sensitive to H_2R-mediated mitogenic signals rather than to an H_1R-mediated inhibition. This theory was supported to some extent by similar observations indicating that IL-1 combined with other mitogenic cytokine signals was able to enhance bone marrow proliferation, although alone it fails to induce hematopoiesis.[10] Interpretation of this data and other available reports regarding histamine-induced hematopoietic proliferation has led to the notion that it was heavily questionable whether histamine should be considered as a common general regulator of bone marrow proliferation. Histamine-provoked proliferation of hematopoietic stem cells is apparently a rather exceptional, intermittent phenomenon limited to periods characterized by an activated immune system, as almost all of the cytokine signals that induce histamine synthesis in the bone marrow are more or less coupled with ongoing immune responses.

Fibroblasts and Fibrosis

Enhanced proliferation, migratory activity, altered production of and interactions with extracellular matrix components (matrix remodeling) are characteristic features of fibroblasts participating in fibrosis. There is a clear and accepted connection between chronic inflammation and fibrotic aberration of connective tissues (e.g., Mycobacteria-infected tissues show heavy granulomatous fibrosis because of permanent inflammation due to unsuccessful immunological clearance of the bacteria), however the fact that mast cell activation could also result in similar consequences is not widely recognized. Since the late 1970s, experimental models have provided strong evidence for abnormal accumulation and enhanced degranulation of mast cells in lung fibrosis[11] and, furthermore, for increased histamine content in the bronchoalveolar lavage of patients suffering from these disorders.[12] As mast cell over-activation appeared in response to many structurally unrelated fibrosis-inducing agents in several experimental settings and the exhibited mitogenic effect of mast cell mediators was not limited to the lung,[13] it was proposed that histamine might be a general inductor of fibrosis.

In their pioneer work Jordana et al[14] first demonstrated in a pure in vitro setting that histamine was in fact able to enhance fibroblast proliferation, an effect that could be abrogated by H_2R but not by H_1R antagonist treatment.[14] More recently a significant advance was made in this area as it was reported that histamine induced both proliferation (via both H_1R and H_2R) and migration of conjunctival fibroblasts (via H_2R dependent signaling pathways).[15] Histamine seemed also to be capable of stimulating procollagen I production in these cells, however this effect remained unaffected by H_1R or H_2R antagonists[15] suggesting involvement of another type of histamine receptor. Finally, similar results were obtained in a study conducted with human fibroblasts treated with sonicated human mast cells, as it was again shown that mast cell components such as histamine and tryptase support fibroblast proliferation and collagen synthesis.[16]

Mammary Gland and Uterus

In the murine uterus rapidly dividing epithelial cells of the endometrium are major sources of histamine. In these cells the level of HDC-expression is controlled mainly by progesterone-mediated signals, which, interestingly, induce maximal level of HDC-expression on the day of implantation (day four of pregnancy). Epithelial cells from the mammary glands of mice, rats, pigs, guinea pigs, cows and humans express HDC at high levels leading to a local histamine concentration that is one of the highest among mammalian tissues.[17] In the mammary glands activity of the HDC gene is under the control of estradiol. Estradiol signal transduction appears to influence HDC expression at the transcriptional level. Estradiol induces HDC expression resulting in a progressively increasing rate of histamine synthesis during the estrus cycle and, in mice and rats, a significant increase in the histamine content of the mammary gland can be observed during the late, proestrus-estrus phase of the cycle.

Moreover, during the development of the mammary glands in young females there is an elevated histamine level along with rapid cell proliferation.[17] Mammary glands are able to synthesize and inactivate histamine, since the activity of both histidine decarboxylase and at least one of the histamine-catabolizing enzymes has been demonstrated in these tissues. Furthermore, expression of both H_1R and H_2R has been detected in the mammary gland.[18] These data suggest that histamine fulfills the basic criteria for involvement in the physiological regulation of mammary tissue. Interestingly, during the development of the mammary glands H_1R and H_2R change the coupling to their respective signal transductional pathways.[19] Between 30 and 60 days of age, as a consequence of progressive branching, the gland of young virgin rats exhibits a profuse mitotic and DNA-synthesizing activity, which declines progressively with age. During this period histamine H_1R and H_2R are coupled to their second messenger systems in a rather uncommon way initially described in mammary adenocarcinoma cells.[20] Consequentially, expression of a high affinity H_2R

that activates phospholipase C (PLC) instead of adenylyl cyclase makes it possible for histamine to modulate cell growth. Later in the adult female tissue, the coupling of the high affinity H_2R changes to adenylyl cyclase, thus playing a role in differentiation. All these data support the hypothesis that histamine is a mediator of hormonal actions regulating proliferation in tissues such as mammary glands and the uterus.

Skin

The ability of keratinocytes to synthesize, store and release histamine has been proposed to be related to the physiological functions of histamine in the skin.[21] Initially these functions were thought to be related to inflammation and itching, however, more recently it was suggested that they might also be coupled to keratinocyte proliferation, differentiation and function.[22] Histamine is released in the skin from both mast cells and keratinocytes; both cell types contain considerable amounts of HDC. The effects of histamine in the skin are mainly mediated by the activation of H_1R.[23] The expression of H_2R on keratinocytes was also reported, but its physiological importance is less clear. Recent observations have suggested that the activation of this receptor may have a role beyond the usually described response to inflammation. Accumulating evidence indicates that the activation of the H_2R may not always increase cAMP production in keratinocytes, but under some circumstances can be connected to the PLC signaling pathway. Interestingly, histamine metabolism is reduced during keratinocyte differentiation and is associated with a significant reduction in HDC mRNA levels, intracellular histamine content and histamine release. This reduction is followed by a substantial down-regulation in the expression of H_2R that influences the functional coupling to the PLC signaling pathway, suggesting that histamine may have a paracrine/autocrine role on the regulation of keratinocyte growth.[24] Histamine could also stimulate the production of IL-6 and other cytokines by keratinocytes, thereby enhancing cell proliferation.[23] The regulation of H_2R expression has been demonstrated in other cell lines during cellular differentiation. Furthermore, strong evidence suggests that histamine could participate in epithelial cell differentiation in an autocrine manner. Consequently, the presence of HDC and the accumulation of histamine in the undifferentiated phenotype, point to a broader function for histamine and for the H_2R as mediators of keratinocyte growth.

Tumor Formation—Principles

The development of malignant tumors is a long process lasting years or decades and recent progress has led to a more detailed picture as to the molecular mechanisms involved. Recent information suggests that there are two main phases in the development of clinically manifested tumors—that is the malignant transformation of normal cells and the subsequent tumorous progression of the transformed clones that results in a gradually independence from the mechanisms controlling cellular growth, tissue integrity and homeostasis. We also know that one major driving forces of this process is the random mutations that cells accumulate during their individual life. This effect modifies their genome and can transform healthy cells as well as constantly leading to the formation of newer mutant variants among the already cancerous cells. The other driving force is continuous clone selection, which sorts out the most resistant clones from the numerous competing tumor cell variants. These cells are then able escape from the grab of the anti-tumor mechanisms activated against them. In the long run this process is the microscopic analogue of the gradual adaptation process of different species evolving in their natural environments, i.e., evolutionary development. This has led to the view that the development of tumors is a type of microevolution. One of the capital conclusions of this concept is that in the body among the permanently appearing precancerous cells (cells that have just started transformation) only an insignificantly small proportion can reach a degree of development where they can form clinically relevant tumors. However, the other message of this theory is that cells found in such tumors have come down a long evolutionary road and are well adapted to the environment surrounding them and to its defensive reactions. In other words, cells of clinically manifested tumors can be considered as a peak in tumor microevolution.

It is well known that there are several intrinsic reasons for initiating and stimulating tumor formation and malignant progression. Such intrinsic reasons are the limited chemical stability of DNA, faulty DNA replication in dividing cells, errors of genomic repair and imprinting systems and the DNA damaging effects of reactive free radicals. Furthermore, there are external transforming effects as well. For instance, physical mutagens like ionizing radiation that cause chain breaks in DNA or ultraviolet radiation-induced dimerizing of adjacent pyrimidine bases. There are also chemical mutagens such as alkylating agents, other agents that interchalate into the DNA double chain, various nucleotide-base analogues and so on. Finally, besides the physical and chemical effects, certain biological factors also belong to the group of external carcinogenic factors. For instance, some pathogens, like the retroviruses, infiltrate randomly into the genome and may inactivate crucial genes or transactivate adjacent genes through insertion of aggressive retroviral promoters. In addition, there are certain DNA-viruses that are not built into the genome, but can produce proteins with transforming effects. Other biological factors are genomic parasites that are located at the border of endogenous and exogenous tumorogenic effects, for instance the various transposons that are moving continuously in the human genome as well as other, similar transposable elements.

To counter these effects numerous regulating, correcting, cellular distress and immunological defensive systems are mobilized to maintain the organism's operability. In the majority of cases these systems are able to keep the process of malignant transformation and cancerous progression in check. These systems include the DNA repair systems mentioned above that protect the integrity of the information coded in the genome and the automatically activated "built-in" cell cycle controlling points and tumor suppressors in dividing cells that stop the further division of the cells, if necessary. These can also start the process of programmed cell death, thereby destroying both recently transformed and, though less frequently, progressing cancer cells. There are also regulating systems at the organism level that inhibit the further spread of precancerous cells by controlling the accessibility of growth factors, cytokines and other different regulating molecules. Beyond the direct obstructive effects on the division and survival of transforming cells, these systems can efficiently impede the oxygen and nutriment supply of the given cancerous cells, their invasion into the adjacent tissues and their emigration into the surrounding blood and lymph circulation. Last, but not least, the organism also directs an aggressive machinery, the immune system, against developing tumors. The immune system deploys its commandingly ample weapon arsenal and presents a waxworks of various sorts of cellular death when tracing and destroying cancer cells.

Histamine in Benign and Malignant Tumors

The first evidence reported on the relevance of histamine in tumor development came when increases in histamine content, urinary excretion and HDC activity was observed in tumor bearing mice.[25] Since 1960 many authors have demonstrated that histamine synthesis is significantly increased in tumoral tissues like breast cancer, colon carcinoma, melanoma, lymphomas and leukemia, compared to surrounding normal tissues.[26,27] In experimental tumor models the role of histamine has been more clearly established. In N-nitroso-N-methylurea (NMU)-induced mammary adenocarcinomas in rats histamine was demonstrated as an autocrine growth factor.[28] In vitro studies employing cell lines have demonstrated the expression of H_1R and H_2R and the associated signaling pathways through which histamine may modulate cell proliferation. Direct effects of histamine on tumor cells were found to be primarily autocrine, although paracrine effects of histamine released by tumor cells influencing immune responses or stimulation of angiogenesis has to also be considered.

Proliferation of Malignant Cell Lines

Several recent literature reports have demonstrated the effect of histamine on the proliferation of different cell lines, but the data remain controversial. The decrease in proliferation caused by HDC antisense oligonucleotides indicates considerable functional relevance of histamine synthesis in melanoma growth.[31] The diverse effects of histamine can be explained by the relative abundance of the different receptors and differences in their affinity for histamine. Histamine receptors are

Table 1. Histamine synthesis and the presence of H_1R, H_2R, in different cell lines derived from human neoplasias

Cell Line/References	Histamine Receptors	Effects on Cell Proliferation
Parietal cell[29] HEK-293	H_2R	Histamine stimulates cell growth
Colon carcinoma.[30] C170, LIM2412	H_2R	Histamine stimulates cell proliferation in C170, but neither histamine nor H_2R antagonist affected basal growth of LIM2412
CT-26.[45]	H_2R	Histamine does not effect proliferation
Melanoma[27,31,44,45] A875, Js, DU, BH, NEW WM-35,983, HT-168, M1	H_1R and H_2R	Different effects depending on histamine concentration and cell line
Pancreatic carcinoma.[33,34] PANC-1	H_1R and H_2R	At low concentration histamine increases proliferation while at high doses histamine inhibits cell growth, producing a G1/G0 cell cycle arrest
Promonocytic cell line.[35,36] U937	H_1R and H_2R	Differentiation induced by histamine via H_2R
Human epidermal carcinoma cells A431 and HeLa.[37]	H_1R	Stimulating proliferation of cells that express functional H_1R
Astrocytoma.[38] U373 MG cells	H_1R	H_1R activation stimulates the proliferation
Prostate cancer cell line.[39] DU-145	H_1R	Histamine inhibits the proliferation by H_1R

expressed in multiple malignant cell types, however the effect of histamine depends on the type of the cell, the balance between different receptor subtypes and the downstream effectors that are subsequently activated. For example, histamine influences the growth of human melanoma cells depending on the actual receptor balance, i.e., it increases cell proliferation through H_2R, but inhibits it through H_1R.[27] In addition H_2R obtain functional heterogeneity by activating adenylate cyclase or phospholipase C by either different subclasses of H_2R or by coupling to two different effector G proteins[29,32] (Table 1). Interestingly, most cell lines express HDC and contain high amount of histamine in a diffusible, unstored form. The histamine released to the extracellular medium is in the nanomolar concentration range, thus clearly indicating that this endogenously synthesized histamine can modulate diverse biological responses through the activation of high affinity histamine receptors in an autocrine way.

In-Vivo Experimental Models

Histamine has proven to be crucial for tumor growth and development in various experimental models. If we consider that histamine is a mediator of keratinocyte growth and hormonal action in the mammary gland, it could then be assumed that it plays a role as a promoter in skin and mammary gland carcinogenesis. In the development of different animal tumors such as EMT6 sarcoma in mice, Lewis Lung carcinoma in mice and HTC hepatoma in rats, a parallel increase in ornithine decarboxylase and HDC activities occurs during early stages. The induction of HDC is followed by a rapid increase in tumor histamine content and can lead to increases in tissues distant from the tumor such as spleen or lymphoid nodes.[25] The specific HDC inhibitor, αFMH, reduces histamine levels in some tissues and produces a marked inhibition of tumor growth in different animal models.[40,41] The multistage model of mouse skin carcinogenesis is a useful system to study the sequential changes that normal skin undergoes during its transformation into neoplastic tissue. Studies of histamine receptors in initiated skin of Sencar mice during the promotion-progression period showed a sequential change in the coupling of both H_1R and H_2R to signal transduction pathways.[42] In uninvolved skin H_1R were coupled to inositol phosphate production and H_2R to cAMP production as normal. However, in tumors this was reversed. A simultaneous increase in mast cell numbers, with a homogeneous subepithelial distribution and marked phenotypic changes was observed suggesting that such cells may play a role in the process of carcinogenesis. These findings indicate an atypical association of histamine receptors to second messengers that could enable histamine to behave as a growth factor through H_2R. This alteration seems to be a significant feature for the postulated action of histamine in tumor growth. Strengthening this hypothesis, PDV-C57 cells, a cultured cell line derived from DMBA-transformed mouse keratinocytes and different mice epidermal tumor cell lines exhibit a similar atypical signal transduction linkage of histamine receptors.[43]

In another study our group previously showed that transgenic enhancement of histamine production in B16-F10 melanomas strongly supports tumor growth in C57BL/6 mice.[44] Gene expression profiles (validated by real-time PCR and immunohistochemistry) of transgenic mouse melanomas secreting different amounts of histamine suggested that H_1R activation suppressed RNA-level expression of the tumor suppressor insulin-like growth factor II receptor (IGF-IIR) and the antiangiogenic matrix protein fibulin-5 (FBLN5). Pathway analysis suggested that since plasma membrane-bound IGF-IIR is required to activate matrix-bound, latent transforming growth factor-B1, a factor suggested to sustain FBLN5 expression, the effect of H_1R activation can be explained by the suppression of a known antineoplastic regulatory pathway. Surprisingly, these data show that in melanoma H_2R are rather irrelevant compared with H_1R.[45]

A critical role has been described for histamine in the development of an experimental mammary adenocarcinoma induced in rats with NMU.[46] In these tumors histamine exerts a regulatory function on cell growth by acting directly through H_1R and H_2R, which exhibit an atypical coupling to intracellular signaling systems. In these tumors histamine acts as an autocrine growth factor and stimulates cell proliferation. Interestingly, in the NMU-induced tumors, HDC loses its normal response to estrogen. Furthermore, HDC mRNA abundance and enzyme activity are regulated by histamine itself acting upon its specific membrane receptors. Therefore, the escape of HDC from the characteristic estrogen control present in the normal mammary gland is a common feature of all NMU-induced tumors and it may be one the first and most significant alterations in the development of the malignant phenotype. The oral administration of ranitidine at 50 mg/kg to rats bearing mammary adenocarcinomas resulted in the complete remission in 55% of tumors, partial remission in 35% while 10% remained without changes. The survival of treated animals was significantly higher than control groups. On the contrary, the treatment with an H_1R antagonist augmented tumor growth rates in all cases and animal survival was significantly decreased. The daily administration of αFMH, 5 mg/kg, produced the remission of 100% of the tumors.[46]

In an experimental syngenic tumor model using a colon adenocarcinoma cell line, CT-26, in Balb/c mice, daily administration of cimetidine significantly suppressed the increase in tumor volume and weight observed on day 6 after inoculation. Cimetidine also reversed the suppression of the expression of tissue cytokines such as lymphotoxin-β, TNF-α, IFN-γ, IL-10 and IL-15. The effects of cimetidine on tumor growth in this model might be mediated by the restoration of local cytokine expression and thereby exert antitumor effects.[47] In human melanoma-grafted mice the combined treatment of cimetidine and DPPE reduced tumor growth and increased survival. These changes were accompanied by enhanced infiltration of IFNγ producing mouse macrophages into the tumor tissue.[48]

Experimental results obtained working with a model of human bronchogenic carcinomas propagated as subcutaneous xenografts in immunosuppressed mice, indicated that the treatment with the H_2R antagonists cimetidine and ranitidine increased the tumor mass doubling time. The co-administration of the dual H_2R/histamine H_4 receptor (H_4R) agonist 4-methyl histamine abolished the effects of cimetidine, thus indicating that in these tumors histamine may stimulate growth through the H_2R.

Human Tumors

In addition to the data available from studies in tumor cell lines derived from human neoplasias and in xenotransplanted models in nude mice, a number of investigations have also been carried out in biopsies of tumoral tissues. Most reports are coincident in showing an increase in histamine synthesis and content in tumors and the presence of histamine receptors.

Breast and Colorectal Cancer

Tumor and adjacent normal breast content of histamine was measured in 29 patients having surgery for breast cancer.[50] The median content of histamine in breast cancer tissue was 5.4 (range 0.9-27.3) μg/g and was significantly greater than that in adjacent breast tissue. The authors postulated that the concentration of histamine in breast cancer was sufficient to inhibit lymphocyte function suggesting that it could be locally immunosuppressive.[50] In agreement with this result, Garcia-Caballero et al reported that the HDC activity of breast cancer is significantly higher than that of the healthy mammary gland tissue of the same patient.[26] It is interesting that the activity of HDC was the highest in benign tumor tissue. It was also found that the histamine content in muscle tissue was significantly higher in cancer patients than in noncancer patients. These findings suggest that high histamine synthesis by malignant tumors can affect other host tissues.[26]

A study with tissue samples obtained by surgery from 25 patients indicated that both H_1R and H_2R are present in human mammary glands, in benign lesion and in breast carcinomas.[51] In benign lesions H_2R produced increases in cAMP levels, while H_1R were coupled to PLC activation. On the other hand, the response observed in carcinomas was different: H_1R were invariably linked to the PLC pathway, but H_2R stimulated both signal transduction pathways. In agreement with a previous report, approximately 25% of breast carcinomas were negative for H_2R. In all cases the lack of expression of H_2R in tumors corresponded to patients with poor prognosis and progression.[51]

Histamine has been found to stimulate growth of colorectal cancer in vitro and in vivo. Similarly to the studies carried out in breast carcinomas, the histamine content was measured in 31 colorectal cancer specimens using a radioenzymatic assay.[52] The median histamine concentration in colorectal cancer tissue was 8.4 μg/g, ranging from 0.3 μg/g to 20.6 μg/g. Again the high concentration of histamine in colon cancer seems high enough to be locally immunosuppressive.[52] Another study was designed to determine the activity of HDC in normal and tumor tissues in a series of ten surgical patients with colorectal carcinoma. A significantly increased HDC activity, almost double that of normal tissues, was found in specimens of human tumors, consistent with numerous reports of high HDC activity in tumor-bearing animals.[53]

In humans, several clinical trials have been carried out with H_2R antagonists as an adjunct to surgical resection, with conflicting results. While the H_2R antagonist cimetidine has been shown to reduce the number of tumor infiltrating lymphocytes in colorectal cancer this was not found to be the case in breast cancer.[54] A comparison of breast cancer with colorectal cancer showed no relationship between preoperative cimetidine administration and tumor cell proliferation. Furthermore, cimetidine did not have any influence on tissue histamine content and mast cell numbers. The presence of mast cells around tumor tissue raises questions concerning the source of histamine in breast tumor tissue. Tumor cell proliferation correlated well with other prognostic indicators such as grade and differentiation.[54]

The beneficial effect of cimetidine on survival in colorectal cancer patients has been demonstrated in several cases, but the mode of action has not been elucidated. In one animal study cimetidine was demonstrated to block the adhesion of colorectal tumor cells and metastasis of the tumor cell in a nude mouse model.[55]

A recent study suggested that in human colorectal cancer tumors both H_1R and H_4R expression was markedly decreased ($p < 0.001$) both at the mRNA and protein levels compared to those of normal colonic mucosa, without significant change in H_2R.[56]

The up-regulation of HDC protein expression and activity in colorectal tumor specimens has been detected and shown to be higher in metastatic tumors than in nonmetastatic ones. These variables significantly correlated with tumor PGE_2 production. These data showed that histamine exerts both a proproliferative and a proangiogenic effect via H_2R/H_4R activation. These effects are likely to be mediated by increasing COX-2-related PGE_2 production in colon cancer cells that express COX-2.[57]

Melanoma

Analysis of HDC expression in melanoma showed that in primary melanomas and cutaneous metastases HDC is present at significantly higher levels than in benign nevi and the surrounding nontumoral skin. Parallel expression of HDC and synthesis of histamine has also been demonstrated in melanoma. Moreover, HDC and the released histamine both have been shown to be related to the malignant phenotype and it is also likely that the presence of these markers may have a prognostic value.[31]

It has been demonstrated that melanomas not only have a histamine-producing capacity, but also perform continuous histamine secretion into the environment as well.[27] In addition, melanoma cells are also able to detect the presence of histamine since surface expression of H_1R and H_2R has been demonstrated. Moreover they most likely also display H_3R, which is in line with the fact that melanocytes originate from the neuronal crest during embryonic development. However, in spite of the above data, the many in vitro observations and the data derived from mouse tumor models, there is no clear evidence that histamine-mediated autocrine signaling has an actual role in the in vivo progression of human melanomas and if it does, which aspects of malignant development are affected. Although definite in vivo evidence for this concept is not yet available, it has been shown that histamine-mediated signals may support the proliferation of melanoma cells in culture.[31] In addition to this it is notable that histamine can efficiently influence the immune system in the local milleu of the tumor and it may also be able to do the same at the antigen-presenting level in draining lymph nodes. Furthermore, it seems that histamine primarily mediates Th2-type effects and, hence, can inhibit anti-tumor Th1 responses. It is also known that in dermal mast cells UVB-radiation, one of the most widely suspected causative agents of melanoma, evokes a robust histamine release that has significant immune-modulating effects frequently even leading to systematic immune-suppression. Observations suggesting that in the stroma of certain tumors, including melanoma, a robust mast cells inflow is perceptible make an interesting parallel with these thoughts and emphasize the possible existence of significant external histamine-release that can also assist melanoma growth. Notably, it has recently become evident that in addition to histamine, several other mediators and enzymes released from mast cells can stimulate invasivity and angiogenesis by increasing capillary permeability and inducing remodeling of the adjacent stroma.[27,31]

Lung Cancer

It was recently reported that human small cell lung carcinomas (SCLCs) express vesicular monoamine transporters in addition to other neuroendocrine markers indicating that SCLCs are histaminergic. The biosynthetic enzyme HDC was detected by immunohistochemistry in paraffin sections of 12 biopsies of SCLC tumors. This finding was supported by immunoblotting and RT-PCR using established SCLC cell lines, as well as, frozen and paraffin-embedded SCLC tumors. Moreover, it was found that histamine is synthesized, stored and released by cultured SCLC cells. These observations may be useful for developing new diagnostic tools for this frequent and highly malignant tumor.[58] There are also data about human lung H_1R and H_2R in cancer and chronic inflammatory processes. It has been found that the number of H_1R significantly increases both in cancer and chronic pneumonia, although it does not change in tuberculosis lung parenchyma. In contrast, the binding parameters of H_2R both in cancer and inflammatory processes were similar to those obtained for the normal tissue. The authors proposed an important role of parenchymal H_1R in the neuromodulation of airways in human lung adenocarcinoma.[59]

Lymphomas and Leukemia

The median levels of histamine were determined in lymph nodes of patients with malignant lymphomas, Hodgkin's disease (HD) or non-Hodgkin's lymphomas (NHL) and in all cases the values were higher than in controls. In patients with NHL, the levels of histamine showed a dependence on the grade of malignancy, as they were significantly higher in those classified as high-grade malignant.[60] Immunostaining and ELISA also confirmed the presence of histamine in the cytoplasm of acute lymphocytic leukemia (ALL) cells. Primary leukemia cells had detectable levels of histamine ranging from 12.5 pg/10^6 cells to 1235.4 pg/10^6 cells. ALL cells can therefore produce histamine and H_1R antihistamines can inhibit their clonogenic growth. However, there was no correlation between the clonogenic growth of ALL cells and their histamine content suggesting that, while histamine may be important for the clonogenic growth of ALL cells, other factors also affect their clonogenicity.[61]

Histamine Blood Levels in Cancer Patients

A number of different authors have demonstrated that histamine serum levels are significantly decreased in cancer patients, suggesting that decreased levels may be a good marker for the onset and progression of solid malignant tumors. In patients with gastrointestinal tumors, blood histamine levels provided information additional to that derived from serum carcinogen embrionary antigen (CEA) determination. In patients with nongastrointestinal tumors, the blood histamine level may be of more value than CEA as a marker of disease progression.[63] In a similar study, the concentration of histamine in blood was determined in 22 patients with solid malignant tumors, 16 hospitalized noncancer patients and 9 healthy subjects. Patients with cancer (mainly carcinomas of the breast and gastrointestinal tract) were divided into two groups: patients with resected primary tumor without known metastases (group I) and patients with present primary tumor with or without metastases (group II). In comparison to healthy subjects (histamine concentration 69.0 ± 6.0 ng/ml), cancer patients in both groups had significantly decreased levels of histamine in blood. Also, the concentration of histamine in patients with present tumor (group II; 40.1 ± 3.48 ng/ml) was significantly lower than in patients with resected primary tumor (49.9 ± 3.14 ng/ml). Furthermore, hospitalized noncancer patients had slightly lower (not significant) concentrations of histamine (59.7 ± 6.13 ng/ml) compared to healthy subjects'.[63] Blood histamine levels were reported to be significantly decreased in patients with active phase Hodgkin's lymphoma and returned to normal values after treatment or during remission. A recent work investigated the utility of a histamine assay for detecting the presence of primary cancers. In order to assess the usefulness of this assay in primary tumor monitoring, two groups of individuals, 29 controls and 29 colon cancer patients were selected and serum levels of histamine, CEA and tumor staging were determined. A significant reduction in histamine levels was found between controls and patients. No correlation was found, however, between tumor node metastasis staging and histamine levels, indicating that this marker is not related to the tumor mass and that it could be a potentially interesting nonspecific tumor marker in colon cancer monitoring.[64]

Angiogenesis

Clinically tumor progression, invasion and metastasis are more relevant than tumor induction. It is well known that the outcome of tumor invasion and metastasis is closely dependent on the capability to induce angiogenesis. Angiogenesis is tightly regulated by pro- and anti-angiogenic factors. Activated mast cells are able to induce and enhance angiogenesis via multiple interacting pathways. In tumor models, mast cells have been shown to play a decisive role in inducing the angiogenic switch, which precedes malignant transformation. In fact, it is well documented that heparin, combined with a range of heparin-binding factors such as bFGF (basic fibroblast growth factor) or TGFβ (transforming growth factor) is able to promote neovascularisation and that mast cell proteases cause cell structural alterations by destroying extracellular matrix integrity. Moreover, there is strong evidence that mast cells significantly influence angiogenesis and thus growth and progression in human cancers.[65] The role of histamine secreted by mast cells is less clear. It seems that histamine can behave as a pro- or an anti-angiogenic modulator, depending on which histamine receptor is bound.[66,67]

Many years ago it has been reported that histamine increases endothelial cell proliferation via both H_1R and H_2R. Histamine is able to induce VEGF production in the granulation tissue via the H_2R-cyclic AMP-protein kinase A pathway and augments angiogenesis in the granulation tissue.[68] Recently it has been demonstrated that HDC-deficient (HDC-KO) mice show lower VEGF levels in the granulation tissue and that there is notably less angiogenesis and granulation tissue formation than in the wild-type mice. The topical injection of histamine or the H_2R agonist dimaprit rescued the defective angiogenesis and granulation tissue formation in HDC-KO mice. This effect is not observed in mast cell deficient mice. In addition, macrophages in the granulation tissue were found to express HDC. These findings indicate that histamine derived from cells other than mast cells play a significant role in the angiogenesis of inflammatory granulation tissue.[69] IP-10, (interferon-induced protein of 10 kDa) induces antitumor immune responses in synergy with IL-12 via activated Th1 cells. In addition, IP-10 inhibits tumor angiogenesis and abrogates a route for nutrition and metastasis. Tumor cells can produce IP-10 in response to IFNγ and histamine inhibits this effect.[70] In different melanoma and squamous carcinoma cell lines, it was observed that histamine suppresses IP-10 expression and secretion via H_2R and the cAMP-protein kinase A signaling pathway.[70] By this mechanism histamine released from tumor cells or peritumoral mast cells may increase angiogenesis and sustain the growth and survival of the tumor.[70]

Effect of Histamine on Immune Regulation

Histamine strongly affects the balance of cytokines from T helper type 1 (Th1) cells and T helper type 2 (Th2) cells by shifting cytokine production from a Th1 to a Th2 pattern. Stress mediators, as well as histamine and adenosine, are increased in certain circumstances and thus upregulate Th2 cytokines and may play a role in the induction and progression of certain allergic/atopic reactions and tumor growth.[27,71] Using a CT-26 colon adenocarcinoma-based syngenic mouse experimental tumor model, endogenous tumor histamine synthesis was shown to enhance tumor growth by suppressing local expression of LT-β (lymphotoxin), TNF-α (tumor necrosis factor alpha) and IFN-γ (interferon gamma) in an H_2R-dependent manner.[73]

Conclusion

Convincing evidence is available that if production of inflammatory mediators is prolonged or their normal functions become modified, they can provide support for both tumor formation and progression. Histamine seems to be one of the modulators in normal cell proliferation in many tissues, including skin, gastric mucosa, bone marrow and fibroblasts. Moreover, there is an increase in histamine synthesis and content in tumors and in most cases the expression levels of histamine receptors are also increased. Histamine serum levels are significantly decreased in cancer patients, suggesting that decreased levels may be a good marker for the onset and progression of solid malignant tumors, Histamine regulates angiogenesis, as well, and it seems that it can behave as a pro- or an anti-angiogenic modulator, depending on which histamine receptor is bound.

References

1. Brenna E, Tielemans Y, Kleveland PM et al. Effect of the histamine-2 agonist impromidine on stem cell proliferation of rat oxyntic mucosa. Scand J Gastroenterol 1995; 30:311-4.
2. Morini G, Grandi D, Schunack W. Ligands for histamine H(3) receptors modulate cell proliferation and migration in rat oxyntic mucosa. Br J Pharmacol 2002; 137:237-44.
3. Chen G, Kashiwagi H, Omura N et al. Effect of a histamine H_1 receptor antagonist on gastric endocrine cell proliferation induced by chronic acid suppression in rats. J Gastroenterol 2000; 35:742-7.
4. Modlin IM, Zhu Z, Tang LH et al. Evidence for a regulatory role for histamine in gastric enterochromaffin-like cell proliferation induced by hypergastrinemia. Digestion 1996; 57:310-21.
5. Ciacci C, Zarrilli R, Ricci V et al. Histamine H_2-receptor antagonists stimulate proliferation but not migration of human gastric mucosal cells in-vitro. Dig Dis Sci 1996; 41:972-8.
6. Byron JW. Mechanism for histamine H_2-receptor induced cell-cycle changes in the bone marrow stem cell. Agents Actions 1977; 7:209-13.

7. Nakaya N, Tasaka K. The influence of histamine on precursors of granulocytic leukocytes in murine bone marrow. Life Sci 1988; 42:999-1010.
8. Endo Y, Kikuchi T, Takeda Y et al. GM-CSF and G-CSF stimulate the synthesis of histamine and putrescine in the hematopoietic organs in-vivo. Immunol Lett 1992; 33:9-13.
9. Schneider E, Piquet-Pellorce C, Dy M. New role for histamine in interleukin-3-induced proliferation of hematopoietic stem cells. J Cell Physiol 1990; 143:337-43.
10. Piquet-Pellorce C, Schneider E, Dy M. GM-CSF in association with IL-1 triggers day-8 CFU-S into cell cycle: role of histamine. J Cell Physiol 1991; 149:18-23.
11. Kawanami O, Ferrans VJ, Fulmer JD et al. Ultrastructure of pulmonary mast cells in patients with fibrotic lung disorders. Lab Invest 1979; 40:717-34.
12. Haslam PL, Cromwell O, Dewar A et al. Evidence of increased histamine levels of lung lavage fluids from patients with cryptogenic fibrosing alveolitis. Clin Exp Immunol 1981; 44:587-93.
13. Franzen, Norrby K. Local mitogenic effect of tissue mast cell secretion. Cell Tissue Kinet 1980; 13:635-42.
14. Jordana M, Befus AD, Newhouse MT et al. Effect of histamine on proliferation of normal human adult lung fibroblasts. Thorax 1988; 43:552-8.
15. Leonardi A, Radice M, Fregona IA et al. Histamine effects on conjunctival fibroblasts from patients with vernal conjunctivitis. Exp Eye Res 1999; 68:739-46.
16. Garbuzenko E, Nagler A, Pickholtz D et al. Human mast cells stimulate fibroblast proliferation, collagen synthesis and lattice contraction: a direct role for mast cells in skin fibrosis. Clin Exp Allergy 2002; 32:237-46.
17. Kierska D, Fogel W, Maslinski C. Histamine concentration and metabolism in mouse mammary gland during estrous cycle. Inflamm Res 1997; 46:63-4.
18. Maslinski C, Kierska D, Fogel WA et al. Histamine: its metabolism and localization in mammary gland. Comp Biochem Physiol C 1993; 105:269-73.
19. Paria B, Das N, Das S et al. Histidine decarboxylase gene in the mouse uterus is regulated by progesterone and correlates with uterine differentiation for blastocyst implantation. Endocrinology 1998; 139:3958-66.
20. Davio C, Cricco G, Bergoc RM et al. H_1 and H_2 histamine receptors in experimental carcinomas with an atypical coupling to signal transducers. Biochem Pharmacol 1995; 50:91-6.
21. Malviya R, Morrison AR, Pentland AP. Histamine in human epidermal cells is induced by ultraviolet light injury. J Invest Dermatol 1996; 106:75-9.
22. Maurer M, Optiz M, Henz BM et al. The mast cells product histamine and serotonine stimulate and TNF-alpha inhibits the proliferation of murine epidermal keratinocytes in situ. J Dermatol Sci 1997; 16:79-84.
23. Shinoda S, Kameyoshi Y, Hide M et al. Histamine enhances UVB-induced IL-6 production by human keratinocytes. Arch Dermatol Res 1998; 290:428-38.
24. Fitzsimons C, Engel N, Policastro L et al. Regulation of phospholipase C activation by the number of H_2 recepotrs during Ca^{2+}-induced differentiation of mouse keratinocytes. Biochem Pharmacol 2002; 63:1785-96.
25. Bartholeyns J, Fozard J. Role of histamine in tumor development. Trends Pharmacol Sci 1985; 6:123-5.
26. Garcia-Caballero M, Neugebauer E, Rodriguez F et al. Histamine synthesis and content in benign and malignant breast tumors. Its effects on other host. Surg Oncol 1994; 3:167-73.
27. Falus A, Hegyesi H, Lazar-Molnar E et al. Paracrine and autocrine interactions in melanoma: histamine is a relevant player in local regulation. Trends Immunol 2001; 22:648-52.
28. Cricco GP, Davio CA, Martin G et al. Histamine as an autocrine growth factor in experimental mammary carcinomas. Agents Actions 1994; 43:17-20.
29. Wang LD, Wang M, Todisco A et al. The human histamine H(2) receptor regulates c-jun and c-fos in a differential manner. Am J Physiol Cell Physiol 2000; 278:C1246-55.
30. Adams WJ, Lawson JA, Morris DL. Cimrtidine inhibits in-vivo growth of human colon-cancer and reverses histamine-stimulated in-vitro and in-vitro growth. Gut 1994; 35:1632-6.
31. Hegyesi H, Somlai B, Varga VL et al. Suppression of melanoma cell proliferation by histidine decarboxylase specific antisense oligonucleotides. J Invest Dermatol 2001; 117:151-3.
32. Mitsuhashi M, Mitsuhashi T, Payan DG. Multiple signaling pathways of histamine H_2 receptors. Identification of an H_2 receptor-dependent Ca^{2+} mobilization pathway in human HL-60 promyelocytic leukemia cells. J Biol Chem 1989; 264:18356-62.
33. Cricco G, Martín G, Labombarda F et al. Human pancreatic carcinoma cell line panc-1 and the role of histamine in growth regulation. Inflammation Res 2000; 49:68-9.
34. Martín G, Cricco G, Darvas Z et al. Histamine inhibits proliferation of a pancreatic carcinoma cell line without inducing apoptosis significantly. Inflamm Res 2002; 51:S67-8.

35. Gespach C, Cost H, Abita JP. Histamine H_2 receptor activity during the differentiation of human monocytic-like cell line U937. FEBS Lett 1985; 184:207-13.
36. Shayo C, Davio C, Brodsky A et al. Histamine modulates the expression of c-fos through H_2 receptor on the human monocytic cell line U937. Mol Pharmacol 1997; 51:983-90.
37. Tilly BC, Tertoolen LG, Remorie R et al. Histamine as a growth factor and chemoattractant for human carcinoma and melanoma cells: action through Ca2(+)-mobilizing H_1 receptors. J Cell Biol 1990; 110:1211-5.
38. Hernandez-Angeles A, Soria-Jasso LE, Ortega A et al. Histamine H_1 receptor activation stimulates mitogenesis in human astrocytoma U373 MG cells. J Neurooncol 2001; 55:81-9.
39. Valencia S, Hernandez-Angeles A, Soria-Jasso LE et al. Histamine H(1) receptor activation inhibits the proliferation of human prostatic adenocarcinoma DU-145 cells. Prostate 2001; 48:179-87.
40. Cricco G, Engel N, Croci M et al. Fluoromethylhistidine inhibits tumor growth without producing depletion of endogenous histamine, Inflammation Research 1997; 46:56-9.
41. Bartholeyns J, Bouclier M. Involvement of histamine in growth of mouse and rat tumors: antitumor properties of monofluoromethylhistidine, an enzyme-activated irreversible inhibitor of histidine decarboxylase. Cancer Res 1984; 44:639-45.
42. Fitzsimons C, Molinari B, Davio C et al. Atypical association of histamine receptors to signal transductional pathways during multistage carcinogenesis in mouse skin. Inflammation Research 1997; 46:292-298.
43. Fitzsimons C, Durán H, Labombarda F et al. Histamine receptors signaling in epidermal tumor cell lines with H-ras gene alterations. Inflammation Research 1997; 47:S50-1.
44. Pos Z, Safrany G, Muller K et al. Phenotypic profiling of engineered mouse melanomas with manipulated histamine production identifies histamine H_2 receptor and rho-C as histamine-regulated melanoma progression markers. Cancer Res 2005; 65:4458-66.
45. Pos Z, Wiener Z, Pocza P et al. Histamine Suppresses Fibulin-5 and Insulin-like Growth Factor-II Receptor Expression in Melanoma Cancer Res 2008; 68:1997-2005.
46. Cricco G, Davio C, Bergoc RM et al. Inhibition of tumor growth induced by histamine in-vivo and in-vitro studies. Agents Actions 1993; 38:75-8.
47. Takahashi K, Tanaka S, Ichikawa A. Effect of cimetidine on intratumoural cytokine expression in an experimental tumor. Biochem Biophys Res Commun 2001; 281:1113-9.
48. Szincsak N, Hegyesi H, Hunyadi J et al. Cimetidine and a tamoxifen derivative reduce tumor formation in SCID mice xenotransplanted with a human melanoma cell line. Melanoma Res 2002; 12:231-240.
49. Sheehan PF, Baker T, Tutton PJ et al. Effects of histamine and 5-hydroxytryptamine on the growth rate of xenografted human bronchogenic carcinomas. Clin Exp Pharmacol Physiol 1996; 23:465-71.
50. Reynolds JL, Akhter JA, Magarey CJ et al. Histamine in human breast cancer. Br J Surg 1998; 85:538-41.
51. Davio C, Cricco G, Bergoc R et al. H_1 and H_2 Histamine receptors in human mammary carcinomas. Agents Actions, Special Conference Issue 1993; 38:71-4.
52. Reynolds JL, Akhter J, Adams WJ et al. Histamine content in colorectal cancer. Are there sufficient levels of histamine to affect lymphocyte function? Eur J Surg Oncol 1997; 23:224-7.
53. Garcia-Caballero M, Neugebauer E, Campos R et al. Increased histidine decarboxylase (HDC) activity in human colorectal cancer: results of a study on ten patients. Agents Actions 1988; 23:357-60.
54. Bowrey PF, King J, Magarey C et al. Histamine, mast cells and tumor cell proliferation in breast cancer: does preoperative cimetidine administration have an effect? Br J Cancer 2000; 82:167-70.
55. Kobayashi K, Matsumoto S, Morishima T et al. Cimetidine inhibits cancer cell adhesion to endothelial cells and prevents metastasis by blocking E-selectin expression. Cancer Res 2000; 60:3978-84.
56. Boer K, Helinger E, Helinger A et al. Decreased expression of histamine H_1 and H_4 receptors suggests disturbance of local regulation in human colorectal tumors by histamine. Eur J Cell Biol 2008; 87:227-36.
57. Cianchi F, Cortesini C, Schiavone N et al. The role of cyclooxygenase-2 in mediating the effects of histamine on cell proliferation and vascular endothelial growth factor production in colorectal cancer. Clin Canc Res 2005; 11:6807-15.
58. Graff L, Frungieri M, Zanner R et al. Expression of histidine decarboxylase and synthesis of histamine by human small cell lung carcinoma. Am J Pathol 2002; 160:1561-5.
59. Kondratenko TY, Zacharova IV, Katukov VYu et al. The study of histamine H_1- and H_2-receptors in human lung cancer. Biochem Mol Biol Int 1993; 31:399-404.
60. Belcheva A, Mishkova R. Histamine content in lymph nodes from patients with malignant lymphomas. Inflamm Res 1995; 44:S86.
61. Malaviya R, Uckun FM. Histamine as an autocrine regulator of leukemic cell proliferation. Leuk Lymphoma 2000; 36:367-73.

62. Burtin C, Noirot C, Paupe J et al. Decreased blood histamine levels in patients with solid malignant tumors. Br J Cancer 1983; 47:367-72.
63. Sabolovic D, Dubravcic D, Culo F et al. Histamine levels in the blood in patients with malignant tumors. Lijec Vjesn 1989; 111:185-7.
64. Previati M, Raspadori A, Bertolaso L et al. Determination of histamine in the whole blood of colon cancer patients. J Chromatogr B Analyt Technol Biomed Life Sci 2002; 780:331-9.
65. Norrby K. Mast cells and angiogenesis. APMIS 2002; 110:355-71.
66. Norrby K. Evidence of a dual role of endogenous histamine in angiogenesis. Int J Exp Pathol 1995; 76:87-92.
67. Fraser RA, Simpson JG. Role of mast cells in experimental tumor angiogenesis. In: Development of the Vascular System (Ciba Foundation Symposium 100). London: Pitman, 1983:120-131.
68. Ghosh AK, Hirasawa N, Ohuchi K. Enhancement by histamine of vascular endothelial growth factor production in granulation tissue via H_2 receptors. Br J Pharmacol 2001; 134:1419-28.
69. Ghosh AK, Hirasawa N, Ohtsu H et al. Defective angiogenesis in the inflammatory granulation tissue in histidine decarboxylase-deficient mice but not in mast cell-deficient mice. J Exp Med 2002; 195:973-82.
70. Kanda N, Watanabe S. Histamine Inhibits the Production of Interferon-induced Protein of 10 kDa in Human Squamous Cell Carcinoma and Melanoma. J Invest Dermatol 2002; 119:1411-9.
71. Elenkov IJ, Chrousos GP, Wilder RL. Neuroendocrine regulation of IL-12 and TNF-alpha/IL-10 balance. Clinical implication Ann N Y Acad Sci 2000; 917:94-105.
72. Idzko M, la Sala A, Ferrari D et al. Expression and function of histamine receptors in human monocyte-derived dendritic cells. J Allergy Clin Immunol 2002; 109:839-46.
73. Takahashi K, Tanaka S, Furuta K et al. Histamine H(2) receptor-mediated modulation of local cytokine expression in a experimental tumor. Biochem Biophys Res Commun 2002; 297:1205-10.

CHAPTER 12

The Future Antihistamines: Histamine H_3 and H_4 Receptor Ligands

Fuqu Yu, Pascal Bonaventure and Robin L. Thurmond*

Abstract

The field of histamine research has progressed far from a century ago when the first biological functions of histamine were identified. It is now known that histamine function is mediated by four histamine receptors, which belong to the G-protein-coupled receptor family. While antihistamines that target the first two receptors have enjoyed clinical and commercial success, efforts to find new antihistamines against the histamine H_3 and H_4 receptors are still in the early stages. Here we will review the therapeutic potential of targeting these new histamine receptors.

Introduction

The first identified histamine receptor was the histamine H_1 receptor (H_1R), which was known from a pharmacological standpoint for many years, but was only cloned in 1991.[1] The H_1R is mainly expressed on smooth muscle and endothelial cells where it mediates vasodilation and bronchoconstriction. The term "antihistamine" has been commonly used to refer to antagonists specific for the H_1R, which are used in the treatment of allergic inflammatory reactions.[2] Several H_1R antagonists have been developed into successful drugs, including diphenhydramine, fexofenadine and loratadine. Diphenhydramine represents a "first-generation antihistamine" that crosses the blood-brain barrier and possesses sedative properties.[3] While in most cases this is considered a side-effect, such drugs also have therapeutic utility as sleep aides. The second-generation antihistamines, such as loratadine, do not cross the blood-brain barrier and this largely alleviates the sedative side effects.[3]

The histamine H_2 receptor (H_2R) was first discovered due to actions of histamine that were not blocked by the existing antihistamines that targeted the H_1R, leading to the recognition that a second histamine receptor existed. The H_2R was actually the first histamine receptor to be cloned.[4] The H_2R is expressed in a variety of tissues including brain, gastric cells and cardiac tissue and its best understood function is its role as a mediator of gastric acid secretion.[5] Together with acetylcholine (M3) and gastrin (CCK_B) receptors, the H_2R is expressed on the basolateral membrane of parietal cells. Stimulation of these receptors leads to movement of H^+, K^+-ATPase to the apical membrane of the cells where it can exchange H^+ for K^+ and results in the secretion of acid. H_2R antagonists, such as cimetidine and ranitidine, are used for the treatment of acid-related diseases like erosive esophagitis, although it has been shown that they are not as effective as proton pump inhibitors, which inhibit the final step of acid secretion.[6]

Given the clinical and commercial success of drugs targeting the H_1R and H_2R, there is much hope for a next generation of antihistamines that specifically target the two newer members of the histamine receptor family, the histamine H_3 (H_3R) and H_4 receptors (H_4R). The H_3R was

*Corresponding Author: Robin L. Thurmond—Johnson & Johnson Pharmaceutical Research & Development, LLC, 3210 Merryfield Row, San Diego, California 92121, USA.
Email: rthurmon@its.jnj.com

Histamine in Inflammation, edited by Robin L. Thurmond.

first described in 1983 as a presynaptic autoreceptor that mediates the synthesis of histamine and inhibits its release from histaminergic neurons in rat brain, but the receptor was not cloned until years later.[7,8] Cloning of the H_4R was first reported in 2000 following a genomic approach based on searches using the H_3R sequence.[9,10] The H_3R is mainly expressed in the nervous system and has been shown to participate in various neuronal activities including cognitive function and sleep-wake regulation.[11] The H_4R is mainly expressed in hematopoietic cells and has been associated with allergic inflammatory and immune responses.[2]

The role of H_3R and H_4R antagonists in various neuronal and inflammatory responses has been extensively investigated. The development of potent and selective ligands has greatly facilitated H_3R and H_4R research and these compounds have demonstrated efficacy in numerous preclinical models. The clinical development of H_3R ligands has progressed further than for the H_4R, with several inverse agonists/antagonists already undergoing clinical evaluation for different indications.[12] H_4R research has largely remained in the preclinical stage, but it is anticipated that H_4R antagonists will progress into clinical studies in the near future. In the following sections, the disease areas where these future antihistamines are being investigated will be discussed.

Potential Indications for H_3R Ligands

The H_3R is predominantly expressed in the central nervous system (CNS). Histamine synthesizing neurons originate in the tuberomamillary nucleus and project throughout all major areas of the brain. The H_3R is a presynaptic autoreceptor on these neurons and is expressed in many brain areas including the cerebral cortex, hippocampus, amygdala, nucleus accumbens, globus pallidus, striatum and hypothalamus.[13,14] In addition, H_3R expression has been identified in nonhistamine-containing neurons in the central and peripheral nervous systems and regulates the levels of a variety of neurotransmitters including noradrenaline and acetylcholine.[13]

Both histamine and the neurotransmitters regulated by it participate in many important physiological functions. Brain histamine levels regulate wakefulness; acetylcholine plays an important role in cognitive functions; and norepinephrine has come to be recognized as playing a large role in attention and focus. Since the H_3R regulates the levels of these important neuronal agents, it has become an attractive target for developing treatments for a variety of neurological disorders (Table 1).

Early H_3R antagonist compounds possess an imidazole moiety that was first considered to be an essential element required for receptor affinity. Examples include thioperamide, ciproxifan and clobenpropit. However, these compounds have not advanced in clinical development as therapeutic agents due to issues such as the potential for drug-drug interactions and poor bioavailability, however they still remain valuable research tools.[15] In addition, some of these compounds have questionable selectivity for the H_3R. For example thioperamide was later found to be a dual H_3R/H_4R antagonist and to have affinity for serotonin receptors and clobenpropit was found to be an H_4R agonist as well as an H_3R antagonist. This has led to the development of non-imidazole-based H_3R antagonists that are more promising clinical candidates.[15] A representative compound in this category is Bioprojet's H_3R inverse agonist/antagonist tiprolisant (BF2.649).[16]

Sleep/Wake Disorders

Narcolepsy is a rare sleep disorder characterized by excessive daytime sleepiness (EDS), either on its own or accompanied by manifestation of cataplexy (sudden loss of muscle tone).[17] The pathophysiology of narcolepsy is strongly associated with an early loss of orexinergic neurons in the hypothalamus liberating hypocretin, an excitatory neurotransmitter promoting wakefulness.[17] Experiments have shown that the histaminergic system is one of the important executive pathways driven by the hypocretin system for promoting wakefulness, indicating that EDS associated with narcolepsy may be due to reduced histamine tone.[18,19] H_3R blockade results in enhanced brain histamine levels, thus counteracting the sleepiness associated with the disrupted hypocretin system observed in narcolepsy. It has been demonstrated that the H_3R inverse agonist/antagonist tiprolisant enhanced neuronal activity in narcoleptic orexin-deficient mice and promoted wakefulness

Table 1. Potential therapeutic indications for histamine H_3 receptor antagonists

Category	Indications	Rationale
Sleep/Wake disorders	Narcolepsy	* Hypocretin is an excitatory neurotransmitter promoting wakefulness * H_3R blockade results in enhanced brain histamine levels, which enhances the activity of the hypocretin system * A general pattern of wake promoting activity of H_3R antagonists has been observed in several species including mice, rats, cats and guinea pigs * Clinical efficacy of tiprolisant in reducing excessive daytime sleepiness in narcoleptic and Parkinson's disease patients
Cognitive impairment	Alzheimer's disease and Schizophrenia	* Acetylcholine is an important neurotransmitter regulating cognitive functions * H_3R antagonists increase acetylcholine levels in the brain and improve performance in a diverse range of rodent cognitive paradigms
	Attention-deficit hyperactivity disorder (ADHD)	* Noradrenaline is an important neurotransmitter for attention * H_3R antagonists may increase noradrenaline levels in the cerebral cortex and thereby provide a therapeutic benefit in ADHD * Clinical efficacy of tiprolisant in ADHD
Other indications	Epilepsy	* The H_3R may have some control over the neuronal hyperactivity associated with seizures in epilepsy * H_3R inhibition reduced epileptic symptoms in various animal models
	Obesity	* Central histaminergic signaling has been implicated in the regulation of appetite * H_3R blockade by inverse agonists or in H_3R knockout mice was shown to reduce food intake and body weight, although inconsistent results have also been reported
	Neuropathic pain	* Previous findings have suggested interactions between histamine, histamine receptors and nociceptors in itch and pain sensation * Conflicting results have been reported regarding the role of the H_3R in pain
	Allergic rhinitis	* The blockade of presynaptic H_3R, through inhibiting histamine release, may enhance norepinephrine release leading to vasoconstriction and reduced congestion * Contribution of the H_3R in histamine-induced nasal blockage has been reported in preclinical models * Efficacy reported in a clinical nasal allergen challenge model
	Cancer	* Histamine upregulation and an increase in histamine receptor expression have been reported in various cancers including colorectal cancer, pancreatic cancer and breast cancer

and decreased abnormal direct onsets of rapid eye movement sleep.[20] Clinical data with tiprolisant has shown reduced EDS in narcoleptic patients and in Parkinson's disease patients.[20,21] Additional data were recently released by Bioprojet indicating that the drug was still efficacious after 6 months treatment.[22] Tiprolisant received orphan drug status by the European Medicines Agency (EMEA) in May 2007. Several other drug candidates are in clinical development for narcolepsy.[12]

Cognitive Impairment

There is evidence that the H_3R can function as a negative modulator of neurotransmission. Blockade of centrally localized H_3R by selective H_3R antagonists has been shown to enhance the release of histamine, acetylcholine and norepinephrine which play important roles in cognitive processes. The cognitive-enhancing effects of H_3R antagonists across multiple cognitive domains in a wide number of preclinical cognition models also bolster confidence in this therapeutic approach for the treatment of attention deficit hyperactivity disorder, Alzheimer's disease and schizophrenia.[23] There are reports of several H_3R antagonists in different stages of clinical development for these disorders.[12,14]

Alzheimer's Disease (AD) and Schizophrenia

Several selective H_3R antagonists have been shown to improve performance in a diverse range of rodent cognition paradigms, including object recognition, olfactory recognition, water maze, radial maze and passive avoidance, with the most pronounced effects being observed in models where an age-related or pharmacologically induced cognitive deficit is present.[24,25] These preclinical results have generated considerable interest in the development of H_3R antagonists as novel treatments for cognitive deficits in conditions such as Alzheimer's disease and schizophrenia.

It is estimated that in 2009 there were more than 5.3 million people affected by Alzheimer's disease in the United States alone.[26] Alzheimer's disease is a neurodegenerative disease characterized by progressive patterns of cognitive and functional impairments. In the early stages, the most commonly recognized symptom is memory loss. As the disease advances, symptoms include confusion, irritability and aggression, mood swings, language breakdown, long-term memory loss and the general withdrawal of the sufferer as their senses decline. In animal models, H_3R antagonists have been shown to increase performance in attention and memory tests and prevent the degradation in performance produced by pharmacological agents such as scopolamine, MK-801, or age.[24,27,28] In contrast, agonists of the H_3R generally produce cognitive impairing effects in animal models.[27] Importantly, dense H_3R binding was detected in medial temporal cortex samples from severe cases of Alzheimer's disease, suggesting for the first time that H_3R receptors are preserved in late-stage disease.[29]

Schizophrenia is a neuropsychiatric and mental disorder affecting around 0.4-0.6% of world population.[30] It is characterized by abnormalities in the perception or expression of reality and most commonly manifests as auditory hallucinations, paranoid or bizarre delusions, or disorganized speech and thinking with significant social or occupational dysfunction. The mechanism of schizophrenia is still poorly understood, but the role of H_3R in regulating cognitive functions via histaminergic signals and other neurotransmitters has encouraged interest in investigating H_3R antagonists as a potential treatment for this disease. In several mouse models of schizophrenia tiprolisant showed significant inhibitory activity. It reduced locomotor hyperactivity elicited by methamphetamine or dizolcipine without significantly affecting spontaneous locomotor activity when administered alone. It also abolished the apomorphine-induced deficit in prepulse inhibition.[31] Several other H_3R antagonists, such as ABT-239 and GSK-189254, have also been shown to be effective in rodent schizophrenia models.[24,32] The pro-cognitive effects of tiprolisant in schizophrenic patients are currently being evaluated through the MATRICS program.[33]

Attention-Deficit Hyperactivity Disorder (ADHD)

ADHD is the most commonly studied and diagnosed psychiatric disorder in children with a worldwide prevalence of about 5%.[34] ADHD is characterized by the co-existence of attention problems and hyperactivity.[35] The etiology of ADHD is unknown, however a combination of factors including genetic and environmental influences appear to contribute to the onset of this disease. Enhancement of cerebral cortical noradrenaline levels may be an essential component in the therapeutic effect of ADHD treatments.[36] Because the H_3R controls noradrenaline release, H_3R antagonists may increase its levels in the cerebral cortex and, therefore, be beneficial in ADHD. Several H_3R antagonists have been shown to be effective in a preclinical model of ADHD. Spontaneously hypertensive rat pups are known to exhibit learning deficits compared to normotensive pups. This memory deficit can be corrected by the administration of methylphenidate, a compound clinically used to treat ADHD and by H_3R antagonists.[13] Several H_3R antagonists have been reported to have progressed into clinical trials for the treatment of ADHD.[14] Positive results from a Phase II study in adult ADHD with tiprolisant were recently reported.[22]

Other Indications

Epilepsy, obesity, pain, allergic rhinitis and cancer have also been suggested as potential indications for ligands of the H_3R, but the role of the H_3R in the mechanisms of these disorders is less clear and there have been controversial reports regarding the effect of H_3R antagonists in preclinical studies. However, there is an ongoing interest in investigating H_3R antagonists as potential treatments for these indications.

Epilepsy

Epilepsy is a common chronic neurological disease characterized by recurrent unprovoked seizures affecting about 50 million people worldwide.[37] The seizures are generally associated with neuronal hyperactivity and reflect an imbalance between excitatory glutamatergic signaling and inhibitory GABAergic signaling. H_3R ligands may prove useful in adjusting this balance and H_3R inhibition has been shown to reduce epileptic symptoms in various animal models.[38-40]

Currently pilot clinical data exist for tiprolisant.[41] In this study, the efficacy of tiprolisant was tested using a photosensitivity technique. Photosensitivity, defined as a generalized epileptiform reaction to intermittent photic stimulation outlasting the stimulus train, is found in about 5% of epileptic patients. The result from the 12 patients enrolled in this study showed an ability of tiprolisant to suppress photo paroxysmal response. Other compounds have been tested in animal epileptic models and in the near future additional H_3R antagonists may progress into clinical testing.

Obesity

Obesity has become a growing health concern worldwide and is a major risk factor for diabetes mellitus. One of the characteristics of obesity is excessive caloric intake. The brain receives and processes energy intake information from the periphery and controls appetite through numerous neurotransmitters and hormones including neuropeptide Y, melanocortin, leptin and ghrelin.[42] Central histaminergic signaling has been implicated in the regulation of this process.[12] However, there have been inconsistent findings as to the role of the H_3R in feeding behavior and body weight. Several studies reported the ability of H_3R blockade either by antagonists or in H_3R-deficient mice to reduce food intake and body weight.[43-45] Other studies, however, reported that H_3R blockade increases appetite and body weight in mice, whereas H_3R activation leads to decreases.[42] Despite the conflicting preclinical data, investigation of H_3R antagonists for their therapeutic potential in diet-induced obesity continues.

So far tiprolisant is the only known H_3R inverse agonist/antagonist for which some clinical data related to feeding behavior are available. In a single-center, open-label, placebo-controlled Phase I study to evaluate subjective satiety, tiprolisant was found to counteract the effect of the antipsychotic drug olanzapine, which is known to reduce satiety and induce weight gain in treated patients.[46] Much work is still needed to establish the clinical relationship between H_3R antagonists and obesity/diabetes, but it remains an exciting area of H_3R research.

Neuropathic Pain

Histamine has been implicated in both pain and itch responses and there are complicated interactions between the two pathways, although itch and pain are largely controlled by peripheral and central nervous systems respectively.[47] Findings such as the fact that in patients with neuropathic hyperalgesia cutaneous injection of histamine results in a sensation of pain instead of itch, suggest an interaction between histamine, histamine receptors and nociceptors.[48] The potential therapeutic effect of agents targeting the central histamine receptor H_3R in pain has been investigated in numerous preclinical models and conflicting findings have been reported.[49,50]

Allergic Rhinitis

Allergic rhinitis is a nasal inflammatory disorder characterized by symptoms such as nasal congestion, pruritus, sneezing and rhinorrhea. It is believed that the interaction of antigens with antigen-specific IgE bound to IgE receptors on the surface of nasal mast cells causes the release of pro-inflammatory mediators that generate the symptoms of the disease.[51] Antihistamines that target the H_1R have been used for many years to treat allergic rhinitis and while they are generally effective, the relief of some symptoms, such as congestion, is not complete. This has led to the suggestion that other histamine receptors may be involved. The H_2R is believed to play a minimal role in allergic rhinitis, since H_2R antagonists appear to provide little benefit. However, there has been preclinical evidence for a contribution of H_3R in histamine-induced nasal blockage.[52,53] The blockade of presynaptic H_3R, through inhibiting histamine release, may enhance norepinephrine release leading to vasoconstriction and reduced congestion. A recent clinical study studied the effects of an H_3R antagonist in a nasal allergen challenge model.[54] Here it was shown that the combination of an H_3R antagonist with fexofenadine was more effective than fexofenadine plus pseudoephedrine in reducing the allergen induced nasal symptom including congestion. Several H_3R antagonists including some that target multiple histamine receptors, such as H_1R/H_3R dual antagonists, have been reported to be in clinical development for allergic rhinitis.[14]

Cancer

Histamine plays a role in normal and cancerous cell proliferation. Histamine upregulation and an increase in histamine receptor expression in various cancers including colorectal cancer, pancreatic cancer and breast cancer have been reported. The relationship between histamine receptors and those cancers has been specifically studied for the H_1R, H_3R and H_4R.[55-57] Histamine receptor antagonists, including H_3R antagonists, may represent a new therapeutic approach to treat these malignancies.

Potential Indications for H_4R Ligands

The histamine H_4 receptor (H_4R) is the newest member of the histamine receptor family. The H_4R is expressed mainly on hematopoietic cells, such as mast cells, eosinophils, basophils, dendritic cells and T cells.[58] Recently, the H_4R has also been shown to be expressed in the CNS including the brain and spinal cord and is expressed in the dorsal root ganglion suggesting that like the H_3R it may have utility in CNS disorders.[59] However, most of the evidence has pointed to a role for the H_4R in inflammatory and immune responses. Due to the short history of H_4R research, the clinical development of potential therapeutic agents targeting the H_4R is not as advanced as for other histamine receptors. However, its hematopoietic expression pattern and the increasing preclinical evidence for H_4R as an important immune and inflammatory response modulator have made it an attractive target to develop novel therapeutic approaches for a number of disorders. The role of the H_4R in various conditions and potential indications for future antihistamines targeting the H_4R will be discussed in the following section (Table 2).

Asthma

Asthma is characterized by acute and reversible bronchoconstriction driven by airway inflammation that is often eosinophilic in nature and affects 300 million people worldwide.[60] The disease can be exacerbated and become a chronic condition after repeated acute episodes of inflammation, which are often in response to specific allergens. Activation of mast cells and macrophages in the airway mucosa leads to activation of dendritic cells, the recruitment and

Table 2. Potential therapeutic indications for histamine H_4 receptor antagonists

Category	Indications	Rationale
Airway inflammatory diseases	Asthma	* Histamine is a known airway constrictor and increased histamine levels have been found in airway and plasma following antigen challenge * The H_4R is functionally expressed in many cell types associated with asthma, including mast cells, eosinophils, dendritic cells, fibroblasts and T cells (especially Th2 cells) * H_4R-deficient mice and mice treated with H_4R antagonists exhibited decreased allergic lung inflammation and related cytokine levels
Pruritus	Atopic dermatitis	* Histamine is a well-known mediator of itch and increase in histamine levels has been observed in the skin and plasma of patients with pruritic conditions such as atopic dermatitis or acute and chronic urticaria * The H_4R is expressed in human dermal fibroblasts and inflammatory dendritic epidermal cells * H_4R agonists upregulated the mRNA of IL-31, an important cytokine produced by activated Th2 cells that is involved in atopic dermatitis * H_4R antagonists were shown to inhibit pruritus and dermal inflammation in a model of atopic dermatitis
Autoimmune diseases	Rheumatoid arthritis, psoriasis, ulcerative colitis, multiple sclerosis	* Histamine or histamine metabolites levels as well as mast cell numbers are increased in autoimmune disease patients such as multiple sclerosis and rheumatoid arthritis patients * H_4R expression has been detected in synovial tissues from rheumatoid arthritis and osteoarthritis patients and correlated with disease severity and duration * H_4R antagonists have been shown effective in animal models such as a rat acute colitis model
Other indications	Allergic rhinitis	* There is a significant increase in the expression of the H_4R in human nasal polyp tissue taken from patients with chronic rhinosinusitis * An H_4R antagonist was shown to cause dose-dependent inhibition of nasal symptoms in a mouse allergic rhinitis model
	Pain	* The H_4R has been shown to express in the CNS * H_4R antagonists have been shown to possess antinociceptive activity in a number of animal pain models
	Cancer	* In vitro results have shown that treatment with H_2R/H_4R antagonists prevented cell proliferation and VEGF production in colon cancer cell lines * H_4R expression has been shown in other cancers such as pancreatic cancer and breast cancer and has been implicated in the disease mechanism * Agonists of the H_4R protect cells from toxicity associated with anti-cancer drugs

activation of Th2 cells and the infiltration of eosinophils and neutrophils. Ongoing inflammation also leads to remodeling of the airways, which further impacts lung physiology and function.[2,61] The commonly prescribed drugs for asthma target either the acute bronchoconstriction for symptomatic control of asthma, such as inhaled beta2-adrenoceptor agonists, or the underlying inflammation, such as with leukotriene modifiers and inhaled corticosteroids.

Histamine has been closely associated with the pathophysiology of asthma. Histamine is a known airway constrictor and increased histamine levels have been found in airways and plasma of asthma patients following antigen challenge.[62] Many cell types associated with asthma express histamine receptors, most notably the H_1R, H_2R and H_4R. The role of the H_1R and H_2R in asthma has been extensively studied. Cells that are thought to play a major role in asthma such as eosinophils, T cells, mast cells and smooth muscle cells have all been shown to express both the H_1R and H_2R and these receptors can mediate cytokine and chemokine secretion.[2] The role of the H_1R in asthma has been supported by studies of asthma models in H_1R-deficient mice. However, in these models H_1R antagonists were shown to be effective only when given during sensitization, but not during the antigen challenge phase. Despite the preclinical data, currently H_1R antagonists are not a front-line treatment for asthma and indeed a meta-analysis of clinical trial data indicates that H_1R antagonists are not effective in treating asthma.[63] H_2R antagonists have largely had no efficacy in asthma.[64,65]

The identification of the H_4R has offered new insights into the effect of histamine and histamine receptors in asthma. The H_4R is expressed in many important immune cells involved in the pathophysiology of asthma. For example, mast cells are a main source of histamine in the lung and it has been shown that histamine enhances mast cell chemotaxis via the H_4R.[66] Increased eosinophil numbers are found in asthmatic lungs and the H_4R has been shown to mediate eosinophil chemotaxis.[67] Recent studies suggest that the H_4R can modulate airway allergic responses via their influence on T-cell activation. H_4R-deficient mice and mice treated with H_4R antagonists exhibited decreased allergic lung inflammation, with decreases in Th2 responses, including decreases in IL-4, IL-5, IL-13, IL-6 and IL-17 levels.[68] Most recently the H_4R has been shown to be functionally expressed on human Th2 cells and the expression level is upregulated by IL-4.[69] In contrast to H_1R antagonists, H_4R antagonists were equally effective during the sensitization and the allergen challenge phase of a mouse asthma model.[68] The H_4R may account for effects of histamine that are not blocked by H_1R antagonists in asthmatic responses and, in addition, there may be an interaction between the two receptors. A recent study demonstrated that in an acute mouse asthma model, the H_1R antagonist mepyramine and the H_4R antagonist JNJ 7777120 exhibited synergistic inhibitory effects on eosinophil accumulation in the bronchoalveolar lavage fluid.[70]

The H_4R may have other effects that could contribute to asthma. In mice, the H_4R mediates IL-4 and IFNγ production from invariant NK T cells and such cells have been implicated in the pathogenesis of asthma in humans.[71,72] The H_4R mediates the migration of lung fibroblasts, which are important contributors of lung remodeling and other fibrotic lung disorders. Histamine augmented the migration of human fetal lung fibroblasts induced by fibronectin and this effect could be blocked by the H_4R antagonist JNJ 7777120.[73] There is also evidence that the H_4R is important for migration and recruitment of regulatory T cells to the lung. In a mouse asthma model the H_4R agonist 4-methylhistamine given intratracheally reduced lung inflammation and airway hyperreactivity.[74] This was accompanied by an increase in the number of regulatory T cells in the lung. Furthermore, the agonist specifically induced the migration of regulatory T cells in vitro. These data suggest that the H_4R may be important in the recruitment and possible activation of regulatory T cells to help downregulate inflammatory responses. However, one caveat with the work is that 4-methylhistamine has other activities besides those at the H_4R. Of particular interest is the fact that 4-methylhistamine is also an H_2R agonist and H_2R activation is thought to provide anti-inflammatory effects. Nevertheless, further exploration into the role of the H_4R in regulatory T-cell function is warranted.

Atopic Dermatitis

Atopic dermatitis is an inflammatory, chronically relapsing, noncontagious, pruritic skin disease. Histamine is a well-known mediator of itch in both normal and diseased skin of patients with atopic dermatitis.[75] An increase in histamine levels has been observed in the skin and plasma of patients with pruritic conditions such as atopic dermatitis or acute and chronic urticaria.[2,76-79] Atopic dermatitis is believed to be driven, at least in the early stages, by Th2 cell responses, because lesions show marked T-cell infiltration and these cells predominantly express "classic" Th2 cytokines such as IL-4, IL-5 and IL-13, especially during the acute phase; however, Th1 responses may have a more dominant role in chronic lesions.[80-82] The H_4R was shown to be expressed on CD4 T cells, with a higher expression on Th2 cells than Th1 cells and this expression was further enhanced in T cells from subjects with atopic dermatitis.[69] The H_4R agonist 4-methylhistamine upregulated the mRNA of IL-31 in Th2 cells and peripheral blood mononuclear cells. Furthermore, the up-regulation was higher in cells from atopic dermatitis subjects.[69] IL-31 is an important cytokine produced by activated Th2 cells and is thought to be involved in both inflammation and pruritus in atopic dermatitis. Increased IL-31 levels have been detected in lesions of allergic contact dermatitis and atopic dermatitis and its serum levels correlate with disease severity in atopic dermatitis.[83-86] In addition, the H_4R is also expressed in human dermal fibroblasts and inflammatory dendritic epidermal cells that may play a role in atopic dermatitis.[87,88] Finally, the H_4R agonist clobenpropit was shown to enhance the chemotaxis of dendritic cells through skin in an in vitro assay and this effect was blocked by JNJ 7777120.[89]

In a Th2-cell-mediated mouse skin inflammation model that mimics several of the features of atopic dermatitis, H_4R antagonists were shown to significantly inhibit inflammation as evidenced by reduced edema and a reduction in the number of eosinophils and mast cells in the skin.[90] In addition many inflammatory cytokines and chemokines such as IL-4, MCP-1, MIP-1α, KC, RANTES, IL-1β, GM-CSF and TNFα were all reduced in the skin after H_4R antagonist treatment. All of these mediators have been found to be associated with atopic dermatitis and several of them have been shown to be decreased with successful treatment.[91,92] One important cytokine is IL-4, which is key for the development of Th2 cells that in turn express more IL-4. Such IL-4-producing T cells are known to drive acute atopic dermatitis lesions and also be present in chronic lesions.[93] MCP-1, MIP-1α, KC and RANTES are important chemokines for the migration of dendritic cells, monocytes, T cells and eosinophils that are found in atopic dermatitis lesions.[81,94,95] In addition the pro-inflammatory cytokines and chemokines such as IL-1β, GM-CSF and TNFα can influence activation and function of many of these same cell types. Therefore, inhibition of these inflammatory mediators by an H_4R antagonist would be expected to modify the extent and persistence of the skin inflammation in atopic dermatitis. In further support of this idea, the H_3R/H_4R antagonist thioperamide was shown to reduce acanthosis, which indicated the development of contact dermatitis and the number of inflammatory cells in eczematous lesion in a mouse model.[96]

Pruritus is the most common complaint from subjects suffering from atopic dermatitis. Traditional H_1R antagonists have proven to be effective in the relief of certain pruritic conditions, such as those mediated by IgE and mast cell degranulation in acute and chronic urticaria.[97] However, they are not deemed effective in treating pruritus in atopic dermatitis, even though they are commonly prescribed by physicians as an adjunctive treatment to topical corticosteroids and calcineurin inhibitors. Analysis of existing clinical data has come to the conclusion that antihistamines that target the H_1R are largely ineffective in treating pruritus in atopic dermatitis patients.[98] This suggests that the itch response in atopic dermatitis may be mediated by pruritogens other than histamine.

Recently there has been evidence that the H_4R mediates pruritus in mice, but via a different mechanism from the H_1R and therefore it may play a role in pruritic responses in atopic dermatitis. In mice, histamine and selective histamine H_4R agonists caused scratching responses, which were almost completely abolished in H_4R knockout mice or by pretreatment with the H_4R antagonist

JNJ 7777120.[99-101] Differential roles for the H_1R and H_4R were observed in mouse models where scratching behavior was induced by histamine or substance P. The H_1R antagonist fexofenadine reduced scratching induced by histamine but not by substance P, whereas the H_4R antagonist JNJ 7777120 significantly reduced both histamine- and substance P-induced scratching.[102] In addition, JNJ 7777120 was shown to be effective in reducing hapten-induced scratching behavior and scratching to IgE-mediated mast cell degranulation.[100,103] The same was seen in the mouse model of atopic dermatitis where scratching to the hapten was reduced by H_4R antagonist treatment.[90]

One interesting observation from the mouse atopic dermatitis study was that the H_4R-mediated pruritus was independent of mast cells suggesting that both the source of histamine and the location of the H_4R were not mast cells.[90] The lack of mast cell involvement was also found for scratching induced by compound 48/80 in mice.[100] Compound 48/80 is known to directly activate nerve fibers and it was suggested was that pruritus was induced by directly activating C-afferent fibers in the skin. The action of the H_4R would then be downstream of the initial neuronal response. H_4R expression has been detected in the dorsal root ganglion, spinal cord and brain and either of these locations may represent the site of action for H_4R antagonists to block itch.[59,104] Indeed, there is some indication that the anti-pruritic effects of H_4R antagonists require CNS penetration. Recent evidence indicates that the H_4R does have neuronal activity as it has been shown that activation of the receptor mediates hyperpolarization of mouse somatosensory cortex neurons and excitation of human enteric neurons.[104,105] The data on the central activity of the H_4R along with the fact that H_4R antagonists are effective in blocking scratching induced by a number of different pruritogens indicate that the receptor may be downstream from several pruritus pathways and antagonists may be efficacious even if histamine is not the initial trigger. These findings support the claim that H_4R antagonists have therapeutic utility for treating pruritus in atopic dermatitis and other indications where traditional H_1R antagonists are not effective.

Allergic Rhinitis

As described previously, H_1R antagonists have traditionally been used to treat allergic rhinitis and the exploration of the potential therapeutic potential of H_3R antagonists is ongoing. Recent data have also pointed to a potential involvement of the H_4R in this disease process. There is a significant increase in the levels of both the H_1R and H_4R in human nasal polyp tissue taken from patients with chronic rhinosinusitis when compared to normal nasal mucosa.[106] In a mouse allergic rhinitis model, JNJ 7777120 caused dose-dependent inhibition of nasal symptoms.[107] One H_4R antagonist has been reported to be in preclinical development for allergic rhinitis.[108]

Autoimmune Diseases

Autoimmune diseases are defined by aberrant immune responses directed against self. Representative autoimmune diseases include rheumatoid arthritis, systemic lupus erythematosus and multiple sclerosis. Histamine has been regarded as a mediator of inflammatory responses in these disorders and changes in histamine levels have been observed in various autoimmune diseases. For example, there are increased histamine levels in cerebrospinal fluid of multiple sclerosis patients, plasma and synovial fluid of rheumatoid arthritis and plasma of psoriatic arthritis patients.[109-112] Higher levels of histamine metabolites were found in Crohn's disease and ulcerative colitis and the levels positively correlated with disease severity.[113-115] Mast cells are a major producer of histamine and increased mast cell numbers are found in many autoimmune conditions, such as in the brains of patients with multiple sclerosis and in synovial fluid from rheumatoid arthritis patients.[116-119] Evidence for mast cell degranulation is observed in active regions of ulcerative colitis or Crohn's disease.[120,121] In the experimental autoimmune encephalomyelitis model, a model for multiple sclerosis, mast cell-deficient mice exhibit significantly reduced onset and severity of the disease.[122,123]

Although there is circumstantial evidence for the involvement of histamine in autoimmune diseases, H_1R and H_2R antagonists have shown little efficacy for these disorders. The identification of the H_4R has provided new hope for targeting histamine receptors as an effective treatment for these diseases, given the expression of the H_4R in related immune cells and its role in mediating immune and inflammatory responses. H_4R expression has been detected in synovial tissues from rheumatoid arthritis and osteoarthritis patients and the variations in H_4R expression in human synovial cells have been suggested to reflect the severity and duration of rheumatoid arthritis.[14] Recently, it was shown that mice deficient in the enzyme that synthesizes histamine, histidine decarboxylase, were protected from K/BxN serum-induced arthritis, but that H_1R or H_2R deficiency did not have the same effect.[124] This suggested that other histamine receptors like the H_3R or H_4R may be involved. In an acute colitis model in rats two selective H_4R antagonists reduced the pathological symptoms and neutrophil infiltration supporting a role for the H_4R in colitis.[125] One interesting observation from this study was that treatment with the H_4R antagonists led to a reduction in tissue TNFα levels and anti-TNFα therapy is effective for the treatment of ulcerative colitis in humans. [125,126]

Pain

Recent studies support claims that the H_4R plays a role in pain modulation. The H_4R has been shown to be expressed in the CNS, including the brain, spinal cord and dorsal root ganglia.[59] H_4R antagonists have been shown to possess antinociceptive activity in several models of pain.[127-130] The H_4R antagonist, JNJ 7777120 was as efficacious as diclofenac in an acute carrageenan-induced inflammatory pain model and in a more chronic CFA-induced pain model.[130] Similar results were seen in a model of osteoarthritis where the maximum efficacy was on par with celecoxib and in a model of post-operative pain where the efficacy approached that of morphine.[130] Activity for a number of compounds has been reported in neuropathic pain models including JNJ 7777120 where the efficacy was superior to that of gabapentin in two different models.[129,130] In addition, selective H_4R antagonists significantly reduced paw edema and hyperalgesia provoked by sub-plantar injection of carrageenan in a rat acute inflammation and hyperalgesia model.[128,131] One issue with these studies is that the doses needed for efficacy appear to be quite high relative to the in vitro potency. This, however, may be accounted for by the fact that CNS penetration appears to be required for activity and that so far all of the tested compounds have a limited half-life in vivo. Nevertheless, the fact that several structurally distinct H_4R antagonists have shown activity in these pain models is supportive of an H_4R specific effect.

At present, the mechanism of action in these pain models is not clear, but as for the anti-pruritic effects it has been suggested that the CNS penetration is required for activity in the pain models. The reported expression and activity of the H_4R in the peripheral and central nervous systems may point to potential mechanisms.[59,104,105] Therefore, the H_4R presents a promising target for the treatment of pain, although more work is needed to uncover the mechanism of action.

Cancer

The H_4R, along with the H_3R, has been recently indicated to play a role in cancer. One study showed that the administration of histamine increased COX-2 expression and activity, cell proliferation and VEGF production in the COX-2-positive colon cancer cell lines HT29 and Caco2 and treatment with the H_2R or H_4R antagonists prevented these effects.[132] The H_4R has also been shown to be expressed in other cancers, such as pancreatic cancer and breast cancer and has been implicated in disease mechanisms.[56,57] The H_4R has been shown to mediate cell cycle arrest in hematopoietic progenitor cells induced by growth factors.[133] Agonists of the H_4R protect cells from toxicity associated with anti-cancer drugs and may be useful in reducing side effects during cancer therapy. Although the available data for oncology are still very preliminary and somewhat inconclusive, they support a connection between the H_4R and cancer and future efforts to investigate the therapeutic potential of H_4R ligands in oncology.

Conclusion

Histamine mediates a variety of important biological and physiological functions and this has been underscored by the success in a number of indications for drugs that target the H_1R and H_2R. Now the more recently identified H_3R and H_4R have become new targets for the development of therapeutic approaches for many neuronal, immune and inflammatory diseases. Some of this work has focused on diseases where histamine is known to be involved, but where H_1R and H_2R antagonists have not been sufficiently effective. However, new areas are also being explored. Several H_3R ligands have progressed through the preclinical stage into clinical studies for an expanding range of indications. The H_4R antagonists programs have not appeared to have progressed beyond the preclinical stage as of yet, but an increasing number of studies have shown efficacy of selective H_4R antagonists in a variety of animal disease models. The work on the H_3R and H_4R has added new chapters to the already rich history of histamine research and should set the stage in the near future for yet another generation of clinically and commercially successful antihistamines.

References

1. Yamashita M, Fukui H, Sugama K et al. Expression cloning of a cDNA encoding the bovine histamine H1 receptor. Proc Natl Acad Sci USA 1991; 88(24):11515-11519.
2. Thurmond RL, Gelfand EW, Dunford PJ. The role of histamine H_1 and H_4 receptors in allergic inflammation: the search for new antihistamines. Nat Rev Drug Discov 2008; 7(1):41-53.
3. McQuade RD, Richlan K, Duffy RA et al. In vivo binding properties of nonsedating antihistamines to CNS histamine receptors. Drug Dev Res 1990; 20(3):301-306.
4. Gantz I, Munzert G, Tashiro T et al. Molecular cloning of the human histamine H2 receptor. Biochem Biophys Res Commun 1991; 178(3):1386-1392.
5. Hill SJ, Ganellin CR, Timmerman H et al. International union of pharmacology. XIII. Classification of histamine receptors. Pharmacol Rev 1997; 49(3):253-278.
6. Mossner J, Caca K. Developments in the inhibition of gastric acid secretion. Eur J Clin Invest 2005; 35(8):469-475.
7. Arrang JM, Garbarg M, Schwartz JC. Autoinhibition of brain histamine release mediated by a novel class (H3) of histamine receptor. Nature 1983; 302(5911):832-837.
8. Lovenberg TW, Roland BL, Wilson SJ et al. Cloning and functional expression of the human histamine H3 receptor. Mol Pharmacol 1999; 55(6):1101-1107.
9. Oda T, Morikawa N, Saito Y et al. Molecular cloning and characterization of a novel type of histamine receptor preferentially expressed in leukocytes. J Biol Chem 2000; 275(47):36781-36786.
10. Liu C, Ma X-J, Jiang X et al. Cloning and pharmacological characterization of a fourth histamine receptor (H_4) expressed in bone marrow. Mol Pharmacol 2001; 59(3):420-426.
11. Stocking EM, Letavic MA. Histamine H3 antagonists as wake-promoting and pro-cognitive agents. Curr Top Med Chem 2008; 8(11):988-1002.
12. Sander K, Kottke T, Stark H. Histamine H3 receptor antagonists go to clinics. Biol Pharm Bull 2008; 31(12):2163-2181.
13. Bonaventure P, Letavic M, Dugovic C et al. Histamine H3 receptor antagonists: from target identification to drug leads. Biochem Pharmacol 2007; 73(8):1084-1096.
14. Tiligada E, Zampeli E, Sander K et al. Histamine H3 and H4 receptors as novel drug targets. Expert Opin Investig Drugs 2009; 18(10):1519-1531.
15. Zhang M, Ballard ME, Pan L et al. Lack of cataleptogenic potentiation with non-imidazole H3 receptor antagonists reveals potential drug-drug interactions between imidazole-based H3 receptor antagonists and antipsychotic drugs. Brain Res 2005; 1045(1-2):142-149.
16. Ligneau X, Perrin D, Landais L et al. BF2.649 [1-{3-[3-(4-chlorophenyl)propoxy]propyl}piperidine, hydrochloride], a nonimidazole inverse agonist/antagonist at the human histamine H3 receptor: preclinical pharmacology. J Pharmacol Exp Ther 2007; 320(1):365-375.
17. Dauvilliers Y, Arnulf I, Mignot E. Narcolepsy with cataplexy. Lancet 2007; 369(9560):499-511.
18. Mignot E, Nishino S. Emerging therapies in narcolepsy-cataplexy. Sleep 2005; 28(6):754-763.
19. Nishino S, Fujiki N, Ripley B et al. Decreased brain histamine content in hypocretin/orexin receptor-2 mutated narcoleptic dogs. Neurosci Lett 2001; 313(3):125-128.
20. Lin JS, Dauvilliers Y, Arnulf I et al. An inverse agonist of the histamine H(3) receptor improves wakefulness in narcolepsy: studies in orexin−/− mice and patients. Neurobiol Dis 2008; 30(1):74-83.
21. Arnulf I. Results of clinical trials of tiprolisant in narcolepsy and Parkinson's disease. S.19.05 Paper presented at: 22nd ECNP Congress, 2009; Istanbul, Turkey.
22. Schwartz JC. Preclinical and clinical pharmacology of H3 receptor inverse agonist. S.19.02 Paper presented at: 22nd ECNP Congress, 2009; Istanbul, Turkey.
23. Esbenshade TA, Browman KE, Bitner RS et al. The histamine H3 receptor: an attractive target for the treatment of cognitive disorders. Br J Pharmacol 2008; 154(6):1166-1181.
24. Medhurst AD, Atkins AR, Beresford IJ et al. GSK189254, a novel H3 receptor antagonist that binds to histamine H3 receptors in Alzheimer's disease brain and improves cognitive performance in preclinical models. J Pharmacol Exp Ther 2007; 321(3):1032-1045.

25. Medhurst AD, Briggs MA, Bruton G et al. Structurally novel histamine H3 receptor antagonists GSK207040 and GSK334429 improve scopolamine-induced memory impairment and capsaicin-induced secondary allodynia in rats. Biochem Pharmacol 2007; 73(8):1182-1194.
26. Alzheimer's Association. 2009 Alzheimer's disease facts and figures. Alzheimers Dement 2009; 5(3):234-270.
27. Witkin JM, Nelson DL. Selective histamine H3 receptor antagonists for treatment of cognitive deficiencies and other disorders of the central nervous system. Pharmacol Ther 2004; 103(1):1-20.
28. Hancock AA, Fox GB. Perspectives on cognitive domains, H3 receptor ligands and neurological disease. Expert Opin Investig Drugs 2004; 13(10):1237-1248.
29. Medhurst AD, Roberts JC, Lee J et al. Characterization of histamine H3 receptors in Alzheimer's Disease brain and amyloid over-expressing TASTPM mice. Br J Pharmacol 2009; 157(1):130-138.
30. Bhugra D. The global prevalence of schizophrenia. PLoS Med 2005; 2(5):e151.
31. Ligneau X, Perrin D, Landais L et al. BF2.649 [1-{3-[3-(4-Chlorophenyl)propoxy]propyl}piperidine, hydrochloride], a nonimidazole inverse agonist/antagonist at the human histamine H3 receptor: Preclinical pharmacology. J Pharmacol Exp Ther 2007; 320(1):365-375.
32. Fox GB, Esbenshade TA, Pan JB et al. Selective H3 receptor (H3R) blockade: broad efficacy in cognition and schizophrenia. Inflamm Res 2005; 54(Suppl 1):S23-24.
33. NCT00690274. Study to demonstrate cognitive enhancing effects of BF2.649. www.clinicaltrials.gov.
34. Polanczyk G, de Lima MS, Horta BL et al. The worldwide prevalence of ADHD: A systematic review and metaregression analysis. Am J Psychiatry 2007; 164(6):942-948.
35. Biederman J, Faraone SV, Taylor A et al. Diagnostic continuity between child and adolescent ADHD: findings from a longitudinal clinical sample. J Am Acad Child Adolesc Psychiatry 1998; 37(3):305-313.
36. Biederman J, Spencer T. Attention-deficit/hyperactivity disorder (ADHD) as a noradrenergic disorder. Biol Psychiatry 1999; 46(9):1234-1242.
37. Kobau R, Zahran H, Thurman David J et al. Epilepsy surveillance among adults—19 States, Behavioral Risk Factor Surveillance System, 2005. MMWR Surveill Summ 2008; 57(6):1-20.
38. Harada C, Fujii Y, Hirai T et al. Inhibitory effect of iodophenpropit, a selective histamine H3 antagonist, on amygdaloid kindled seizures. Brain Res Bull 2004; 63(2):143-146.
39. Harada C, Hirai T, Fujii Y et al. Intracerebroventricular administration of histamine H3 receptor antagonists decreases seizures in rat models of epilepsia. Methods Find Exp Clin Pharmacol 2004; 26(4):263-270.
40. Jiang X, Chen A, Li H. Histaminergic modulation of excitatory synaptic transmission in the rat basolateral amygdala. Neuroscience 2005; 131(3):691-703.
41. Schwartz J-C, Lecomte J-M, Schwartz J-C et al. (Bioprojet, Fr.). assignee. Treatment of epilepsy with non-imidazole alkylamine histamine H3 receptor ligands. US Patent 2006103537. 20060330, 2006.
42. Yoshimoto R, Miyamoto Y, Shimamura K et al. Therapeutic potential of histamine H3 receptor agonist for the treatment of obesity and diabetes mellitus. Proc Natl Acad Sci USA 2006; 103(37):13866-13871.
43. Masaki T, Yoshimatsu H, Chiba S et al. Central infusion of histamine reduces fat accumulation and upregulates UCP family in leptin-resistant obese mice. Diabetes 2001; 50(2):376-384.
44. Tokita S, Takahashi K, Kotani H. Recent advances in Mol Pharmacol of the histamine systems: physiology and pharmacology of histamine H3 receptor: roles in feeding regulation and therapeutic potential for metabolic disorders. J Pharmacol Sci 2006; 101(1):12-18.
45. Ishizuka T, Hatano K, Murotani T et al. Comparison of the effect of an H(3)-inverse agonist on energy intake and hypothalamic histamine release in normal mice and leptin resistant mice with high fat diet-induced obesity. Behav Brain Res 2008; 188(2):250-254.
46. Raga MM, Sallares J, Guerrero M et al. (Ferrer Internacional, S.A., Spain). assignee. Preparation of 1-[3-[3-(4-chlorophenyl)propoxy]propyl]piperidine monohydrochloride as a histamine H3 receptor ligand. US Patent 2006084833. 20060206, 2006.
47. Stander S, Schmelz M. Chronic itch and pain—similarities and differences. Eur J Pain 2006; 10(5):473-478.
48. Baron R, Schwarz K, Kleinert A et al. Histamine-induced itch converts into pain in neuropathic hyperalgesia. Neuroreport 2001; 12(16):3475-3478.
49. Mobarakeh JI, Takahashi K, Yanai K. Enhanced morphine-induced antinociception in histamine H3 receptor gene knockout mice. Neuropharmacology 2009; 57(4):409-414.
50. Smith FM, Haskelberg H, Tracey DJ et al. Role of histamine H3 and H4 receptors in mechanical hyperalgesia following peripheral nerve injury. Neuroimmunomodulation 2007; 14(6):317-325.
51. Skoner DP. Allergic rhinitis: definition, epidemiology, pathophysiology, detection and diagnosis. J Allergy Clin Immunol 2001; 108(1 Suppl):S2-8.

52. Taylor-Clark T, Sodha R, Warner B et al. Histamine receptors that influence blockage of the normal human nasal airway. Br J Pharmacol 2005; 144(6):867-874.
53. McLeod RL, Mingo GG, Herczku C et al. Combined histamine H1 and H3 receptor blockade produces nasal decongestion in an experimental model of nasal congestion. Am J Rhinol 1999; 13(5):391-399.
54. Romero FA, Allan RJ, Phillips PG et al. The Effects of an H_3 Receptor Antagonist in a Nasal Allergen Challenge Model. J Allergy Clin Immunol 2010; 125(2):AB191.
55. Masini E, Fabbroni V, Giannini L et al. Histamine and histidine decarboxylase up-regulation in colorectal cancer: correlation with tumor stage. Inflamm Res 2005; 54(Suppl 1):S80-81.
56. Medina V, Croci M, Crescenti E et al. The role of histamine in human mammary carcinogenesis: H3 and H4 receptors as potential therapeutic targets for breast cancer treatment. Cancer Biol Ther 2008; 7(1):28-35.
57. Cricco GP, Mohamad NA, Sambuco LA et al. Histamine regulates pancreatic carcinoma cell growth through H3 and H4 receptors. Inflamm Res 2008; 57(Suppl 1):S23-24.
58. Huang J-F, Thurmond R. The new biology of histamine receptors. Curr Allergy Asthma Rep 2008; 8(1):21-27.
59. Strakhova MI, Nikkel AL, Manelli AM et al. Localization of histamine H4 receptors in the central nervous system of human and rat. Brain Research 2009; 1250(Complete):41-48.
60. Bateman ED, Hurd SS, Barnes PJ et al. Global strategy for asthma management and prevention: GINA executive summary. Eur Respir J 2008; 31(1):143-178.
61. Barnes PJ. New drugs for asthma. Nat Rev Drug Discov 2004; 3(10):831-844.
62. Busse WW, Swenson CA. The relationship between plasma histamine concentrations and bronchial obstruction to antigen challenge in allergic rhinitis. J Allergy Clin Immunol 1989; 84(5 Pt 1):658-666.
63. Van Ganse E, Kaufman L, Derde MP et al. Effects of antihistamines in adult asthma: a meta-analysis of clinical trials. Eur Respir J 1997; 10(10):2216-2224.
64. Leopold JD, Hartley JP, Smith AP. Effects of oral H1 and H2 receptor antagonists in asthma. British J Clin Pharmacol 1979; 8(3):249-251.
65. Nogrady SG, Hahn AG. H2-receptor blockade and exercise-induced asthma. Br J Clin Pharmacol 1984; 18(5):795-797.
66. Hofstra CL, Desai PJ, Thurmond RL et al. Histamine H_4 receptor mediates chemotaxis and calcium mobilization of mast cells. J Pharmacol Exp Ther 2003; 305(3):1212-1221.
67. Ling P, Ngo K, Nguyen S et al. Histamine H_4 receptor mediates eosinophil chemotaxis with cell shape change and adhesion molecule upregulation. Br J Pharmacol 2004; 142(1):161-171.
68. Dunford PJ, O'Donnell N, Riley JP et al. The histamine H_4 receptor mediates allergic airway inflammation by regulating the activation of CD4+ T-cells. J Immunol 2006; 176(11):7062-7070.
69. Gutzmer R, Mommert S, Gschwandtner M et al. The histamine H4 receptor is functionally expressed on TH2 cells. J Allergy Clin Immunol 2009; 123(3):619-625.
70. Deml K-F, Beermann S, Neumann D et al. Interactions of histamine H1-receptor agonists and antagonists with the human histamine H4-receptor. Mol Pharmacol 2009; 76(5):1019-1030.
71. Pichavant M, Matangkasombut P, DeKruyff RH et al. Natural killer T-cells regulate the development of asthma. Expert Rev Clin Immunol 2009; 5(3):251-260.
72. Leite-de-Moraes MC, Diem S, Michel M-L et al. Cutting Edge: Histamine Receptor H4 Activation Positively Regulates In Vivo IL-4 and IFN-gamma Production by Invariant NKT Cells. J Immunol 2009; 182(3):1233-1236.
73. Kohyama T, Yamauchi Y, Takizawa H et al. Histamine stimulates human lung fibroblast migration. Mol Cell Biochem 2010; 337(1-2):77-81.
74. Morgan RK, McAllister B, Cross L et al. Histamine 4 receptor activation induces recruitment of FoxP3+ T-cells and inhibits allergic asthma in a murine model. J Immunol 2007; 178(12):8081-8089.
75. Steinhoff M, Neisius U, Ikoma A et al. Proteinase-activated receptor-2 mediates itch: a novel pathway for pruritus in human skin. Journal of Neuroscience 2003; 23(15):6176-6180.
76. Johnson HH Jr, DeOreo GA, Lascheid WP et al. Skin histamine levels in chronic atopic dermatitis. J Invest Dermatol 1960; 34:237-238.
77. Juhlin L. Localization and content of histamine in normal and diseased skin. Acta Dermato-Venereologica 1967; 47(6):383-391.
78. Greaves MW, Soendergaard J. Urticaria pigmentosa and factitious urticaria. Arch Dermatol 1970; 101(4):418-425.
79. Kaplan AP, Horakova Z, Katz SI. Assessment of tissue fluid histamine levels in patients with urticaria. J Allergy Clin Immunol 1978; 61(6):350-354.
80. Akdis CA, Akdis M, Bieber T, et al. Diagnosis and treatment of atopic dermatitis in children and adults: European Academy of Allergology and Clinical Immunology/American Academy of Allergy, Asthma and Immunology/PRACTALL Consensus Report. The Journal of Allergy and Clinical Immunology. 2006;118(1):152-169.

81. Homey B, Steinhoff M, Ruzicka T et al. Cytokines and chemokines orchestrate atopic skin inflammation. J Allergy Clin Immunol 2006; 118(1):178-189.
82. Bieber T. Atopic Dermatitis. New England Journal of Medicine 2008; 358(14):1483-1494.
83. Raap U, Wichmann K, Bruder M et al. Correlation of IL-31 serum levels with severity of atopic dermatitis. J Allergy Clin Immunol 2008; 122(2):421-423.
84. Sonkoly E, Muller A, Lauerma AI et al. IL-31: a new link between T-cells and pruritus in atopic skin inflammation. J Allergy Clin Immunol 2006; 117(2):411-417.
85. Gambichler T, Kreuter A, Tomi NS et al. Gene expression of cytokines in atopic eczema before and after ultraviolet A1 phototherapy. Br J Dermatol 2008; 158(5):1117-1120.
86. Neis MM, Peters B, Dreuw A et al. Enhanced expression levels of IL-31 correlate with IL-4 and IL-13 in atopic and allergic contact dermatitis. J Allergy Clin Immunol 2006; 118(4):930-937.
87. Ikawa Y, Shiba K, Ohki E et al. Comparative study of histamine H4 receptor expression in human dermal fibroblasts. J Toxicol Sci 2008; 33(4):503-508.
88. Dijkstra D, Stark H, Chazot PL et al. Human Inflammatory Dendritic Epidermal Cells Express a Functional Histamine H_4 Receptor. J Invest Dermatol 2008; 128(7):1696-1703.
89. Baumer W, Wendorff S, Gutzmer R et al. Histamine H4 receptors modulate dendritic cell migration through skin—immunomodulatory role of histamine. Allergy 2008; 63(10):1387-1394.
90. Cowden JM, Zhang M, Dunford PJ et al. The Histamine H4 Receptor Mediates Inflammation and Pruritus in Th2-Dependent Dermal Inflammation. J Invest Dermatol 2010; 130(4):1023-1033.
91. Park CW, Lee BH, Han HJ et al. Tacrolimus decreases the expression of eotaxin, CCR3, RANTES and interleukin-5 in atopic dermatitis. Br J Dermatol 2005; 152(6):1173-1181.
92. Hatano Y, Katagiri K, Takayasu S. Increased levels in vivo of mRNAs for IL-8 and macrophage inflammatory protein-1a (MIP-1a), but not of RANTES mRNA in peripheral blood mononuclear cells of patients with atopic dermatitis (AD). Clinical and Experimental Immunology 1999; 117(2):237-243.
93. Hamid Q, Boguniewicz M, Leung DYM. Differential in situ cytokine gene expression in acute versus chronic atopic dermatitis. J Clin Invest 1994; 94(2):870-876.
94. Akdis CA, Akdis M, Bieber T et al. Diagnosis and treatment of atopic dermatitis in children and adults: European Academy of Allergology and Clinical Immunology/American Academy of Allergy, Asthma and Immunology/PRACTALL Consensus Report. The J Allergy Clin Immunol 2006; 118(1):152-169.
95. Avgerinou G, Goules AV, Stavropoulos PG et al. Atopic dermatitis: new immunologic aspects. Int J Dermatol 2008; 47(3):219-224.
96. Seike M, Furuya K. Thioperamide (H3 and H4 receptor antagonist) improves ecezematous lesions in murine contact dermatitis. Sagami Joshi Daigaku Kiyo Shizenkei 2007; 71B:7-13.
97. Kaplan AP. Chronic urticaria: Pathogenesis and treatment. J Allergy Clin Immunol 2004; 114(3):465-474.
98. Klein PA, Clark RAF. An evidence-based review of the efficacy of antihistamines in relieving pruritus in atopic dermatitis. Arch Dermatol 1999; 135(12):1522-1525.
99. Yu F, Wolin RL, Wei J et al. Pharmacological characterization of oxime agonists of the histamine H4 receptor. J Receptor Ligand Channel Res 2010; 3:37-49.
100. Dunford PJ, Williams KN, Desai PJ et al. Histamine H_4 receptor antagonists are superior to traditional antihistamines in the attenuation of experimental pruritus. J Allergy Clin Immunol 2007; 119(1):176-183.
101. Bell JK, McQueen DS, Rees JL. Involvement of histamine H_4 and H_1 receptors in scratching induced by histamine receptor agonists in BalbC mice. Br J Pharmacol 2004; 142:374-380.
102. Yamaura K, Oda M, Suwa E et al. Expression of histamine H4 receptor in human epidermal tissues and attenuation of experimental pruritus using H4 receptor antagonist. J Toxicol Sci 2009; 34(4):427-431.
103. Rossbach K, Wendorff S, Sander K et al. Histamine H4 receptor antagonism reduces hapten-induced scratching behaviour but not inflammation. Experimental Dermatology 2009; 18(1):57-63.
104. Connelly WM, Shenton FC, Lethbridge N et al. The histamine H4 receptor is functionally expressed on neurons in the mammalian CNS. Br J Pharmacol 2009; 157(1):55-63.
105. Breunig E, Michel K, Zeller F et al. Histamine excites neurones in the human submucous plexus through activation of H1, H2, H3 and H4 receptors. J Physiol (Oxford, United Kingdom) 2007; 583(2):731-742.
106. Jokuti A, Hellinger E, Hellinger A et al. Histamine H4 receptor expression is elevated in human nasal polyp tissue. Cell Biol Int 2007; 31(11):1367-1370.
107. Takahashi Y, Kagawa Y, Izawa K et al. Effect of histamine H4 receptor antagonist on allergic rhinitis in mice. International Immunopharmacology 2009; 9(6):734-738.
108. Lock R, Collingwood S, Ratcliffe A. Emerging therapies for respiratory diseases. Drug News Perspect 2007; 20(9):593-600.

109. Tuomisto L, Kilpelainen H, Riekkinen P. Histamine and histamine-N-methyltransferase in the CSF of patients with multiple sclerosis. Agents and Actions 1983; 13(2-3):255-257.
110. Petersen LJ, Hansen U, Kristensen JK et al. Studies on mast cells and histamine release in psoriasis: the effect of ranitidine. Acta Dermato-Venereologica 1998; 78(3):190-193.
111. Frewin DB, Cleland LG, Jonsson JR et al. Histamine levels in human synovial fluid. J Rheumatol 1986; 13(1):13-14.
112. Zhang M, Venable JD, Thurmond RL. The histamine H_4 receptor in autoimmune disease. Expert Opin Investig Drugs 2006; 15(11):1443-1452.
113. Winterkamp S, Weidenhiller M, Otte P et al. Urinary excretion of N-methylhistamine as a marker of disease activity in inflammatory bowel disease. Am J Gastroenterol 2002; 97(12):3071-3077.
114. Bischoff SC, Grabowsky J, Manns MP. Quantification of inflammatory mediators in stool samples of patients with inflammatory bowel disorders and controls. Dig Dis Sci 1997; 42(2):394-403.
115. Raithel M, Matek M, Baenkler HW et al. Mucosal histamine content and histamine secretion in Crohn's disease, ulcerative colitis and allergic enteropathy. Int Arch Allergy Immunol 1995; 108(2):127-133.
116. Ibrahim MZM, Reder AT, Lawand R et al. The mast cells of the multiple sclerosis brain. Journal of Neuroimmunology 1996; 70(2):131-138.
117. Toms R, Weiner HL, Johnson D. Identification of IgE-positive cells and mast cells in frozen sections of multiple sclerosis brains. Journal of Neuroimmunology 1990; 30(2-3):169-177.
118. Crisp AJ. Mast cells in rheumatoid arthritis. J R Soc Med 1984; 77(6):450-451.
119. Godfrey HP, Ilardi C, Engber W et al. Quantitation of human synovial mast cells in rheumatoid arthritis and other rheumatic diseases. Arthritis and Rheumatism 1984; 27(8):852-856.
120. Dvorak AM, McLeod RS, Onderdonk A et al. Ultrastructural evidence for piecemeal and anaphylactic degranulation of human gut mucosal mast cells in vivo. Int Arch Allergy Immunol 1992; 99(1):74-83.
121. King T, Biddle W, Bhatia P et al. Colonic mucosal mast cell distribution at line of demarcation of active ulcerative colitis. Dig Dis Sci 1992; 37(4):490-495.
122. Secor VH, Secor WE, Gutekunst C-A et al. Mast cells are essential for early onset and severe disease in a murine model of multiple sclerosis. J Exp Med 2000; 191(5):813-821.
123. Brown MA, Tanzola MB, Robbie-Ryan M. Mechanisms underlying mast cell influence on EAE disease course. Molecular Immunology 2002; 38(16-18):1373-1378.
124. Rajasekaran N, Solomon S, Watanabe T et al. Histidine decarboxylase but not histamine receptor 1 or 2 deficiency protects from K/BxN serum-induced arthritis. Int Immunol 2009; 21(11):1263-1268.
125. Varga C, Horvath K, Berko A et al. Inhibitory effects of histamine H_4 receptor antagonists on experimental colitis in the rat. Eur J Pharmacol 2005; 522(1-3):130-138.
126. Van Assche G, Vermeire S, Rutgeerts P. Infliximab therapy for patients with inflammatory bowel disease: 10years on. Eur J Pharmacol 2009; 623(Suppl.):S17-S25.
127. Altenbach RJ, Adair RM, Bettencourt BM et al. Structure-Activity Studies on a Series of a 2-Aminopyrimidine-Containing Histamine H4 Receptor Ligands. J Med Chem 2008; 51(20):6571-6580.
128. Coruzzi G, Adami M, Guaita E et al. Anti-inflammatory and antinociceptive effects of the selective histamine H_4-receptor antagonists JNJ7777120 and VUF6002 in a rat model of carrageenan-induced acute inflammation. Eur J Pharmacol 2007; 563(1-3):240-244.
129. Cowart MD, Altenbach RJ, Liu H et al. Rotationally Constrained 2,4-Diamino-5,6-disubstituted Pyrimidines: A New Class of Histamine H4 Receptor Antagonists with Improved Druglikeness and in Vivo Efficacy in Pain and Inflammation Models. J Med Chem 2008; 51(20):6547-6557.
130. Hsieh GC, Chandran P, Salyers AK et al. H4 receptor antagonism exhibits anti-nociceptive effects in inflammatory and neuropathic pain models in rats. Pharmacol Biochem Behav 2010; 95(1):41-50.
131. Liu H, Altenbach RJ, Carr TL et al. cis-4-(Piperazin-1-yl)-5,6,7a,8,9,10,11,11a-octahydro benzofuro[2,3-h]quinazolin-2-amine (A-987306), A New Histamine H4R Antagonist that Blocks Pain Responses against Carrageenan-Induced Hyperalgesia. J Med Chem 2008; 51(22):7094-7098.
132. Cianchi F, Cortesini C, Schiavone N et al. The role of cyclooxygenase-2 in mediating the effects of histamine on cell proliferation and vascular endothelial growth factor production in colorectal cancer. Clin Cancer Res 2005; 11(19, Pt. 1):6807-6815.
133. Petit-Bertron A-F, Machavoine F, Defresne MP et al. H4 histamine receptors mediate cell cycle arrest in growth factor-induced murine and human hematopoietic progenitor cells. PLoS One 2009; 4(8):e6504.

Index

P

R

S

T

U

V

W

GPSR Compliance

The European Union's (EU) General Product Safety Regulation (GPSR) is a set of rules that requires consumer products to be safe and our obligations to ensure this.

If you have any concerns about our products, you can contact us on ProductSafety@springernature.com

In case Publisher is established outside the EU, the EU authorized representative is:

Springer Nature Customer Service Center GmbH
Europaplatz 3
69115 Heidelberg, Germany